Lecture Notes in Computer Science 8408

Commenced Publication in 1973
Founding and Former Series Editors:
Gerhard Goos, Juris Hartmanis, and Jan van Leeuwen

T0212676

Jia Li   Wen Gao

Visual Saliency Computation

A Machine Learning Perspective

 Springer

Authors

Jia Li
Peking University, Haidian District, Beijing, 100871, China
E-mail: jia.li@pku.edu.cn

Wen Gao
Peking University, Haidian District, Beijing, 100871, China
E-mail: wgao@pku.edu.cn

ISSN 0302-9743   e-ISSN 1611-3349
ISBN 978-3-319-05641-8   e-ISBN 978-3-319-05642-5
DOI 10.1007/978-3-319-05642-5
Springer Cham Heidelberg New York Dordrecht London

Library of Congress Control Number: 2014934972

LNCS Sublibrary: SL 6 – Image Processing, Computer Vision,
Pattern Recognition, and Graphics

Typesetting: Camera-ready by author, data conversion by Scientific Publishing Services, Chennai, India

Printed on acid-free paper

Springer is part of Springer Science+Business Media (www.springer.com)

Foreword

There are two sides to every story. The same might be said about visual attention. On one hand, due to our limited perceptual and cognitive resources, we have to focus our attention to the most pertinent subset of the available sensory data. On the other hand, we are active seekers of visual information rather than passive recipients, and thus our human vision system often operationalizes attention as a selection mechanism responsible for filtering out unwanted information in a visual scene.

Closely related to attention is the concept of saliency. The saliency of an item - be it an object, a person, a pixel, etc. - arises from its contrast relative to its neighbors. In addition, saliency can be influenced by training: for example, particular letters can become salient for human subjects by training. Saliency detection is considered to be a key attentional mechanism, which guides visual attention. Although attention and saliency have been intensely studied, what we know is still far less than what we do not know. I am, however, optimistic that active investigation in psychology, cognitive neuroscience, and computer science will help reveal more and more about the inner workings of human vision systems.

Visual attention has played an essential role in our biological survival through the development of humanity. It will still play an essential role in this coming Big Data age. As is reported, it will take an individual over half a million years to watch all the videos that currently cross the network each month and the number will increase to 5 million years in 2017. Over 300 million new photos are added to Facebook every day. To "survive" in this era, we have to overcome a number of challenges. How can we efficiently process the massive amount of visual data with limited computational resources? What algorithms do we develop to extract pertinent information to succinctly represent the visual data? How can we model each individual's interest and provide information only relevant to that individual? Visual saliency computation, which measures the importance of various visual subsets in an image or video, is key to addressing these challenges.

This monograph, written by Dr. Jia Li and Prof. Wen Gao, is timely. It uniquely approaches visual saliency computation from a machine learning perspective. Instead of directly simulating the "known" mechanisms of the human brain, they propose to incorporate the modern machine learning algorithms to automatically mine the probable saliency mechanisms from user data. In this process, the prior knowledge, which is believed to be stored in the higher regions of the human brain, can be effectively and efficiently modeled to guide the computation of visual saliency. A multi-task learning technique is developed to infer what a human subject may attend to in an incoming scene by analyzing the users' activities when watching similar scenes in the past. Moreover, they present a statistical learning approach to infer such priors from millions of images in an unsupervised manner. In particular, they have discovered that acquiring the ordering of various visual subsets is more crucial than obtaining their real saliency values. Accordingly, they propose a learning-to-rank approach and formulate the problem of saliency computation within a ranking framework. Subsequently, they examine the properties of salient targets and describe how to extract a salient object from an image as a whole. Extensive experiments are conducted, and the results demonstrate that learning-based approaches have superior performance compared to traditional saliency models.

The book is a delight to read. The authors have put tremendous effort into making it clear and precise. A reader can start from the very basic question of "What is visual saliency?" and progressively explore the problems in modeling saliency, extracting salient objects, mining prior knowledge, evaluating performance, and using saliency in real-world applications. This timely book will be very valuable to researchers and practitioners who are interested in visual information processing and multimedia.

December 2013   Zhengyou Zhang

Acknowledgements

This monograph is based on Dr. Jia Li's doctoral dissertation, under the supervision of Prof. Wen Gao. This book is translated from the dissertation and extended with new content during the two authors' attachment to Peking University.

Many of our colleagues provided us with valueable assistance during the writing of doctor dissertation as well as this edition, which includes valuable material, dataset, suggestions, feedbacks, and comments. We want to thank Dr. Yuanning Li, Dr. Jingjing Yang, Prof. Shuqiang Jiang, Ms. Shu Fang, Ms. Guanghui Cai, Ms. Dan Li and Mr. Xiaobiao Wang for their kind help. A special thank to Prof. Yonghong Tian at Peking University, the leader of the multimedia learning research group at the National Engineering Laboratory on Video Technology at Peking University, who was in the advisory committee on Dr. Li's Ph.D. dissertation, and made many contributions on research and publication. We want to give our thanks to Prof. Fang Fang, Prof. Tiejun Huang and Prof. Yizhou Wang at Peking University, Prof. Qingming Huang at University of Chinese Academy of Sciences, Prof. Xilin Chen at Institute of Computing Technology, Chinese Academy of Sciences, they read the manuscript and gave many comments and suggestions which helped us to improve this book. We are also grateful to Prof. Alex Kot who provided a prosperous environment for writing the book during Dr. Li's visiting to the ROSE lab at Nanyang Technological University, Singapore.

Finally, we appreciate the financial support from the research funds. This work is supported in part by grants from the Chinese National Natural Science Foundation under contract No. 61370113 and No. 61035001, National Basic Research Program of China (No. 2009CB320900) and the Supervisor Award Funding for Excellent Doctoral Dissertation of Beijing (No. 20128000103). This research is also partially supported by the Singapore National Research Foundation under its IDM Futures Funding Initiative and administered by the Interactive & Digital Media Programme Office, Media Development Authority.

Contents

Chapter 1
Introduction

In this chapter, we explain the motivation of visual saliency computation. This chapter introduces the human vision system by showing some basic concepts and evidence from neurobiology and psychology. These pieces of evidence are concluded to support the computational saliency models in this book. At the end of this chapter, we present a structural overview of the book and briefly conclude the main contribution of our research.

1.1 What Is Visual Saliency

Jimmy is a young researcher who enjoys a regular life in the university. Every morning, he "reads" the newspaper when having the breakfast. After that, he drives to the laboratory while "looking" around for safety. During the working hours, he "views" the progress report first to update the daily plan and then "observes" the instruments to get the latest experimental results. After office hours, he prefers to "see" a basketball game and usually ends the day by "watching" TV at home. Jimmy is keeping busy in most time of the day, and so does his vision system.

Actually, 80% of the information we receive every day is from our vision system. In every second, our retina can receive up to 10 billion bits information (Raichle, 2010), while the total number of neurons in cerebral cortex only reaches about 10 billion (Shepherd, 2003) or 20 billion (Koch, 2004). This fact indicates that the amount of visual information we receive remarkably exceeds the storage capability of our brain. Furthermore, the processing capability of our brain is also limited and it is infeasible to simultaneously perform complex analysis on all the input visual information. In order to address these problems, one central task of human vision system is to efficiently detect the *important* visual subsets, say that, the *salient* subsets. These subsets quickly pop-out and get processed with high priorities using the limited

Fig. 1.1 Our vision system receives massive visual information everyday

computational resource in brain, while other subsets are often inhibited or even ignored to increase the processing efficiency.

To further explore the capability of human brain in information processing, we compare it with the fastest and most powerful supercomputers in the world. In November 2013, it is reported that Fujitsu's "K computer" is the 4th powerful supercomputer in the world. However, a new breakthrough in August 2013 reported that Japanese and German scientists utilized 82,944 processors and 1 petabyte of memory on the "K computer" to simulate the activity of 1.73 billion neurons connected by 10.4 trillion synapses, which correspond to only about 10% neurons and 1% connections in human brain. In the simulation, the supercomputer takes about 40 minutes to compute the neuron activities of 1 biological second. Note that a neuron can roughly fire about only 200 times per second, while a computer processer can conduct billions of floating-point operations per second (i.e., several million times faster than a neuron). When feeling proud of the capability of our brain, one question may arise: how can the neural network with billions of slow neurons remarkably outperforms the supercomputer? To this question, one possible explanation is that massive input signals are efficiently processed, highly summarized and sparsely represented at the very early stage before they enter the higher areas of human brain. In particular, it is believed that visual saliency plays an important role in such mechanisms to process the massive visual information received by our vision system.

In the past decades, the field of computer vision, which refers to the techniques on acquiring, processing, analyzing and understanding of the visual world, has invoked many research interests. One objective of developing the *computer vision system* can be described as to electronically duplicate the abilities of *human vision system*. However, the same challenge also arises in this field along with the booming of digital camera and videocorder in the new century. As reported in (Cisco, 2009), Internet videos in 2013 will be nearly 700 times more than those in 2000. In 2013, it will take an individual over half a million years to watch all the videos that cross the network each month and the number will increase to 5 million years in 2017 (Cisco, 2013). When processing the massive visual resource, existing computer vision systems also face a new challenge: how to efficiently process the massive visual information with limited computational resource? Moreover, the processing results should be consistent with human cognition, which is an important objective of the computer vision system.

To solve these problems caused by big visual data, one feasible solution is to learn from the human vision system. That is, we can simulate the mechanism that is used to detect the salient visual subsets in human vision system. With a similar mechanism in the computer vision system, limited computational resource can be allocated to important visual subsets with high priorities for efficient analysis. Moreover, the analysis results can well meet human perception by focusing on the same salient visual subsets as human vision system does. Toward this end, the problem of *visual saliency computation* is proposed, whose objective can be described as predicting, locating and mining the salient visual information by simulating the corresponding mechanisms in human vision system.

In the rest of this Chapter, we will first specify several mechanisms of human vision system that are tightly correlated with visual saliency computation. Following these mechanisms, we will simply introduce how to compute visual saliency and give a structure overview of the book.

1.2 Visual Saliency in Human Vision System

To compute visual saliency, a key step is to build a saliency model by simulating the inherent mechanisms of human vision system. Toward this end, we will first introduce several fundamental concepts that are related with visual saliency. After that, we will demonstrate several finds from neurobiological and psychological studies to reveal the information processing mechanisms in human vision system. Finally, we will summarize the main mechanisms that are important for building saliency models.

1.2.1 Basic Concepts

When talking about saliency, the concept of "attention" should be put into consideration with the topmost priority. Actually, attention and saliency are tightly correlated in both neurobiological structure and psychological definitions. Some researchers in the field of computer vision even use these two terms interchangeably in their works. Thus the first problem is:

•What is attention?

Among all the neurological and cognitive functions, attention is often considered as a very basic function that is often a precursor to all other ones. With the development of modern psychology, the definition of attention is also changing in different periods. In the early research of psychology, Wilhelm Maximilian Wundt, known today as one of the founding figures of modern psychology, proposed that attention should be understood in terms of the differing degrees to which representations are present in consciousness (Wundt, 1893). In (James, 1950), William James characterized attention as the focalization and concentration of consciousness (as shown in Fig. 1.2). In later studies, Alexander Romanovich Luria, one of the founders of cultural-historical psychology, emphasized the selectivity property of attention (Luria, 1973). In (Solso, 1988), Solso proposed that attention means the actively processing of a limited amount of information from all the information available through senses, memories and other cognitive processes.

From the discussions above, we can see that attention can be best referred to as the allocation of cognitive resources on information. Since there are numerous definitions on attention, the second concern arises:

•How many types of attention exist in our brain?

In our daily life, there exist many types of attention that can be well categorized into several major types by their ways to allocate cognitive resources. According to the hierarchic model defined in (Sohlberg and Mateer, 1989), five major types of attention, from the lowest level to the highest level, are defined as:

Everyone knows what attention is. It is the taking possession by the mind, in clear and vivid form, of one out of what seem several simultaneously possible objects or trains of thought. Focalization, concentration, of consciousness are of its essence. It implies withdrawal from some things in order to deal effectively with others, and is a condition which has a real opposite in the confused, dazed, scatterbrained state which in French is called distraction, and Zerstreutheit in German.

-William James, in "The Principles of Psychology (1890)"

Fig. 1.2 Attention defined by William James in (James, 1950)

1. **Focused attention**: the ability to respond *discretely* to specific visual, auditory or tactile stimuli (e.g., read the overlaid captions in a movie).
2. **Sustained attention**: the ability to maintain a *consistent* behavioral response during continuous and repetitive activity (e.g., stay attentive during the movie).
3. **Selective attention**: the ability to *selectively* maintain the behavioral or cognitive resource on specific stimuli while ignoring the distracting or competing stimuli (e.g., focus only on the protagonist in a complex movie scene while ignoring the other people in the same scene).
4. **Alternating attention**: the ability to *switch* between multiple tasks with different demands (e.g., stop watching the movie and read the SMS from a friend).
5. **Divided attention**: the ability to respond *simultaneously* to multiple tasks with different demands (e.g., whisper with friends when watching the movie).

Among all these five types of attention, existing studies in the field of computer vision mainly focus on the selective attention on the visual stimuli, namely, the *selective visual attention*. That is, the studies on computer vision system aim to simulate the mechanisms in human vision system on how to selectively attend to some visual subsets while ignoring the other subsets. In the rest of this book, both the words attention and visual attention, if not specified, indicate the selective visual attention.

From the definition on selective visual attention, we can see that the selectivity is the primary concern. Thus the third problem is:

•**How can visual saliency affect the selectivity?**

Suppose that we are watching a large painting with rich contents, it is difficult to simultaneously perceive the details from all parts of the painting. In this case we may start from the most conspicuous part of the painting and then shift our gaze to other parts after a short period (as shown in Fig. 1.3). For the gaze shifting process, one possible explanation is that each part of the painting has an importance value, or namely, a saliency value. This saliency value can be used as the guide to drive our visual attention to scan across different parts of the painting.

In many scenarios, the selectivity of attention, especially the overt attention, can be characterized as the gaze policy on saliency maps, i.e., the shift of attention between various salient visual subsets. Without loss of generality, we can conclude that visual saliency is a key component to the attentional mechanism by allocating limited perceptual and cognitive resources on the most pertinent subsets of the sensory data. Consequently, we have to address two main issues to compute visual saliency: how to keep or enhance the important visual subsets and how to filter or suppress the unwanted subsets. Thus the last problem arises:

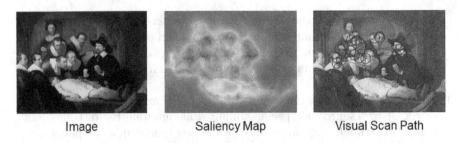

Image   Saliency Map   Visual Scan Path

Fig. 1.3 Human gaze policy is affected by visual saliency. Images are selected from the benchmark proposed in (Le Meur et al, 2006).

•How does the human vision system detect the salient visual subsets?

With this question in mind, we will refer to the latest neurobiological and psychological findings for possible answers. In modern neurobiology and psychology, there exist many tools to explore the information processing mechanisms in human vision system, e.g., single-unit recording, electroencephalography (EEG), magnetoencephalography (MEG) and functional magnetic resonance imaging (fMRI). All these tools can be used to record the information transmission among various components of the human (or monkey) vision system. In some experiments, these devices can be also used to infer the probable functions of specific vision components. Note that the experimental studies of human vision system in neurobiology and psychology are beyond the main concern of this book. To ensure the readability of the book, we only briefly introduce the important findings in these two domains without explaining too many experimental details. People who are interested in the experimental settings could refer to (Kanwisher and Wojciulik, 2000; Baluch and Itti, 2011) and (Itti et al, 2005) for more details.

1.2.2 Neurobiological Mechanisms of Human Vision

In this subsection, we will first introduce the processing mechanisms when visual information is flowing from retina to higher brain regions. During the flowing of information, various visual subsets will compete with each other to become salient. We will also demonstrate how the competitions between visual subsets get biased by the control signals oriented from the higher regions of the brain.

1.2.2.1   Retina

The processing of visual information in our vision system begins in the retina, which is a thin layer of neural tissue in the back of the eye to turn the light into nerve impulses. As shown in Fig. 1.4, the retina is composed of different neural layers (Palmer, 1999), in which the visual information is efficiently preprocessed before entering the brain. For the sake of simplification, the information flow in retina is summarized into three main stages:

1. Photorecepters and horizontal cells. Photorecepters such as cones and rods can turn the light signals into electrical signals in neuron network. Some of these signals are further transmitted to horizontal cells to get the average outputs.
2. Bipolar cells and amacrine cells. Bipolar cells take the difference between the outputs from photorecepters and horizontal cells to get the highly contrasted signals. Amacrine cells provide a second local average of the output from some bipolar cells.
3. Ganglion cells. Ganglion cells are usually situated near the inner surface of the retina. They receive visual information from bipolar cells and amacrine cells and send visual information to different parts of brain along the optic nerve fibres.

As above, two preprocessing mechanisms operate in retina: differencing and averaging. With the two operations, visual information can be effectively compressed and encoded, which is a key function of retina. Note that there are about 128 million photorecepters in the retina, including about 120 million rods and nearly 8 million cones (Sternberg et al, 2008), to capture the light signals. However, there are only about one million optical nerves to transmit

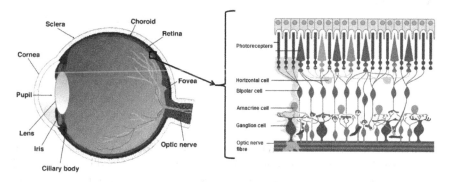

The structure of human eye   The cellular structure of retina

Fig. 1.4 The structure of human eye and the cellular organization of retina. Part of the image is reproduced from (Wilkinson-Berka, 2004) with permission from © Cambridge University Press.

the corresponding neuron responses to the brain. Actually, Raichle (2010) roughly estimated that up to 10 billion bits information per second can arrive at the retina, while only six million bits per second can leave the retina[1]. Therefore, we can safely assume that there exist some structures in retina for data compression using averaging and differencing operations.

Experimental studies show that two center-surround structures have been implemented in retina by the bipolar and ganglion cells, including the on-centers and the off-centers. These structures are not physical but only logical in the sense of the connections between bipolar and ganglion cells. As shown in Fig. 1.5, the on-center cells have an excitory central region and an inhibitory peripheral region, and vice versa for the off-center cells. Such properties can be effective in detecting the local irregularities (e.g., the edges) which are considered to be more informative than the smooth regions. In this manner, the input visual signals can be greatly compressed to fit the limited capacity of the optic nerve fibres.

Logical structure	Stimulus & Response			
	No stimulus	Only on center	Only on peripheral	On both regions
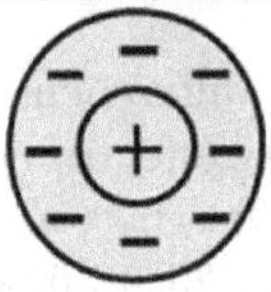 On-center	No response	Strong response	No response	Weak response
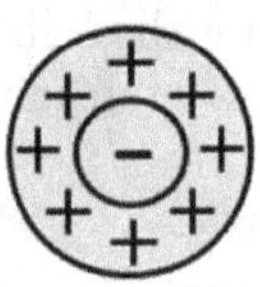 Off-center	No response	No response	Strong response	Weak response

Fig. 1.5 Two center-surround logical structures in retina

1.2.2.2   Lateral Geniculate Nucleus

After the preprocessing in retina cells, the nerve impulses are then sent to the brain in the form of parallel information streams. As concluded in (Baluch and Itti, 2011), most of the information reaches lateral geniculate nucleus (LGN) and only a small portion of information connects to the superior colliculus (SC). When receiving the visual information from the ganglion cells in retina,

---

[1] These numbers are only rough estimations. Readers can refer to (Koch et al, 2006) to obtain a complete analysis of the amount of information transmitted to brain.

two kinds of signals will be generated by the parvocellular cells (i.e., the P-cells) and magnocellular cells (i.e., the M-cells) in LGN, respectively. The small P-cells are sensitive to spatial details, while the big M-cells are capable of resolving motion and coarse outlines. Generally, these two kinds of cells will generate two kinds of nerve responses, in which the output from P-cells has higher spatial resolution and the output from M-cells has higher temporal resolution.

1.2.2.3   Primary Visual Cortex and the Bottom-Up Process

The two outputs from LGN will then feed into the primary visual cortex (V1). Usually, the primary visual cortex is considered to be the start of information processing in the cortical feed-forward visual pathways, including the dorsal and ventral pathways[2]. Existing researches on monkeys show that these two pathways have different functions in processing visual information. The ventral pathway gets its main input from the P-cells, which descends from the primary visual cortex (V1) toward the temporal cortices through extrastriate areas V2 and V4. This stream is believed to be responsible for processing the color, shape, and identity of visual stimuli. Thus it can be described as the "what" pathway. In the dorsal pathway, signals from M-cells ascents from V1 towards parietal cortices through medial temporal (MT) and medial superior temporal (MST). This pathway is responsible for processing location and motion information. Thus it can be described as the "where" pathway.

In the ventral and dorsal pathways, visual information is transmitted from primary visual cortex to some higher brain regions. In the transmission, visual information is gradually summarized and various visual stimuli will compete with each other. In the competition, the winners will gradually pop-out and become salient, while the other stimuli will be suppressed. Since data in this process is transmitted from "bottom (primary)" brain regions to "top (higher)" regions, the process can be identified as the "bottom-up" process and visual saliency generated in this process can be described as the bottom-up saliency. Since the information transmission and competition in this process are mainly affected by data, it can be also described as "data-driven" process. To facilitate the understanding of the bottom-up process, we elucidate the major information processing stages of our vision system in Fig. 1.6.

---

[2] Ventral and dorsal are derived from Latin words. Ventral refers to the bottom surface of the brain, while dorsal refers to the upside of the brain (Sternberg et al, 2008).

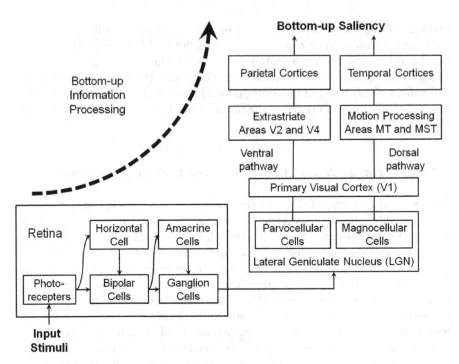

Fig. 1.6 The bottom-up information processing in human vision system

1.2.2.4   The Top-Down Process

Generally speaking, the neurobiological mechanisms of the bottom-up process can well reveal part of the information flow in human vision system. However, our vision system is not a simple logic circuit. One interesting fact is, different neuron responses may occur even for the input signals with exactly the same physical characteristics. A possible explanation for this phenomenon is that our brain is not "empty" when processing the input visual information. It is often fulfilled with tasks, experience and prior knowledge. During the information transmission from "bottom" to "up", these priors in our brain, which are usually stored in higher brain regions, will generate certain control signals from the "top" area to bias the stimuli competition in the "down" area. In this manner, the same visual information will be processed in different ways while using different tasks, experience and prior knowledge, leading to different neuron responses. According to the transmission direction of such control signals, we can call such process as the top-down process, for which we will demonstrate the key findings in neurobiology to clarify three major concerns:

1. How many types of top-down process exist in our brain?
2. Where is the destination of the top-down signal?
3. What is the influence of the top-down signal?

1.2.2.5   Types of Top-Down Processes

When visual information is flowing in the brain, many brain regions (e.g., the parietal-frontal network) can generate top-down control signals to modulate the bottom-up process. Thus an interesting problem may arise: how many types of top-down Processes exist in our brain?

Generally speaking, various top-down modulation processes launched from higher cortical regions can be divided into two categories, including the volitional category and the mandatory category. The volitional top-down process is mainly related with acts of will. For example, when performing a pedestrian searching task, volitional top-down signals will actively bias the competition between visual signals to pop-out only the pedestrians in the scene. This indicates that even the same individual with the same prior knowledge may respond differently to the same stimuli when the tasks are different.

Compared with the volitional process, the mandatory top-down process is automatic, pervasive and cannot be completely eliminated by volition. This process is considered to be related to experience and prior knowledge. Chun and Jiang (1998) reported that our experience from viewed scenes can be automatically transferred to the new scenes with similar spatial layouts (i.e., the contextual modulation of perception). For instance, we often avoid searching cars from the regions outside the road in surveillance videos, while such experience is often from past similar scenes. This fact indicates that the top-down process will always work, even when the subjective willing is absent.

1.2.2.6   Destinations of Top-Down Signals

The top-down control signals originate from higher brain regions and are then transmitted through the volitional or mandatory top-down processes to participate in certain stages of the information flow. In the pure bottom-up process, visual signals will compete without bias to become salient, while the participation of such top-down processes often leads to a biased information selection by favoring a specific category of visual stimuli, spatial locations and feature dimensions. Thus a classic issue is to locate the cortical regions that receive these top-down signals and thus perform biased information selection.

In existing studies, there are two distinct views on when the biased selection will occur (Kanwisher and Wojciulik, 2000), including the early selection view and the late selection view. The early selection view suggests that the top-down signals participate in the early stage of information processing. Only rudimentary perceptual processing is carried out in the very early stage and top-down signals (i.e., focused attention) are necessary for further perceptual analysis such as object identification. In contrast, the late selection view holds that the input information can be analyzed to a high level (e.g., with the identification of objects) before the participation of top-down attentional signals.

With the development of modern physiological tools (e.g., fMRI), it is possible to directly measure the neural response to stimulus. The experimental results strongly support the former view and demonstrate that the top-down modulations can be found at the very early stage of visual processing pathway. It is reported that the top-down modulatory signals are found even in the LGN, which is the first region to receive the signals from retina. Moreover, when the visual information is transmitted along the visual pathways to higher cortical regions, the top-down signals act progressively stronger and sooner. This indicates that the influence of top-down signals may operate on various kinds of low-level (e.g., simple local contrast), mid-level (e.g., scene gist) and high-level features (e.g., complex objects such as face, car and pedestrian).

1.2.2.7   Influence of Top-Down Signal

During the top-down modulations, one of the key concerns is how does the top-down signal affect the neuron response? To answer this question, we have to measure the difference between responses from neurons linked with attended and unattended stimuli (i.e., neurons linked with or without top-down signals). Evidence from single-unit studies on monkey vision system shows that top-down signals can modulate the gain of neural responses. It is believed that top-down signals act by enhancing the responses from neurons that are linked with attended stimuli (e.g., the objective of a visual search task), while the responses for unattended stimuli are inhibited. Actually, the enhancement of targets and inhibition of distractors are two major tasks for both human vision system and computer vision system.

In the literature, there exist two hypotheses on the actual forms in which top-down modulations may take effect, including additive baseline shift and multiplicative gain modulation. The additive hypothesis claims that response to a given stimulus when attended should be higher by a constant than response to the same stimulus when unattended. Strong evidence for this hypothesis is from the experiments which reported increased neuron response when subjects are imagining virtual objects in brain without any external visual input (i.e., only top-down signal is linked with the neuron). The multiplicative hypothesis holds that response to a given stimulus when attended should be the product of a gain multiplier and response to the same stimulus when unattended. Further studies show that these two kinds of influences may be derived from different but partly overlapping neurons. These two hypotheses may both hold, while the precise relationship is unknown. This conclusion is very important for the computational modeling of visual saliency when considering the top-down influence. It determines whether the top-down factors in the saliency model should increase the target saliency and decrease the distractor saliency in an additive framework, or a multiplicative framework, or both.

1.2.3   Psychological Explanations of Visual Saliency

Beyond the neurobiological mechanism, we also show some psychological explanations about the latent mechanism of visual saliency computation in human vision system. Actually, some of the phenomena observed in psychological studies can be well explained by the neurobiological mechanisms discussed above. Here we only introduce the classical psychological phenomena and explanations that are tightly correlated with visual saliency.

1.2.3.1   Pop-Out and Set-Size

Neurobiological studies show that the bottom-up process starts from earlier in visual processing while the top-down process starts from LGN with stronger influences in higher cortical regions. This may imply that the bottom-up process is faster than the top-down process (Wolfe et al, 2000). Actually, two famous psychological phenomena may prove this assumption, including the pop-out and set-size effects.

As shown in Fig. 1.7, pop-out occurs when the target is significantly distinct from the distractors (e.g., several vertical bars among massive horizontal bars). In this case, targets can rapidly pop-out in the bottom-up process and become salient, while the time cost is independent of the distractor number. In contrast, the set-size effect occurs when targets and distractors share certain features (e.g., three vertical green bars among a set of horizontal green/red bars and vertical red bars). In this case, targets cannot efficiently pop-out in the bottom-up process and a much slower top-down modulation should be conducted, making the time cost increase with the number of distractors (i.e., size of the distractor set). Targets can be efficiently distinguished from distractors using simple bottom-up process when they are significantly distinct, and the time cost is irrelevant to distractor number. When targets and distractors share certain features, slow top-down modulation is necessary to find the target and the time cost will increase with the distractor number

Inspired by this fact, we should always model visual saliency by using the bottom-up component as the first module. The top-down component could be added after the bottom-up component, or gradually get involved in the late stages of the bottom-up component.

1.2.3.2   Location, Object or Feature?

When watching a scene, what are the units of our attention? This is a long-standing question in attention research. Actually, our attention may operate on spatial locations, visual features or whole objects. The exact attentive unit is very import to visual saliency computation since it determines whether to detect the salient location, salient feature or salient object.

Many psychological studies have been conducted to address this concern. Most of these studies are carried out by fixing the other two units while changing the rest one, e.g., recording the neuron response in brain areas when subjects attend to different features (e.g., color, shape or motion) of the same visual arrays at the same locations. In some of the experiments (Eriksen and Eriksen, 1974; Connor et al, 1997), attention is directed to specific spatial locations for efficient stimuli selection. However, some other studies (O'Craven et al, 1997; Anllo-Vento et al, 1998; O'Craven et al, 1999) reported that attention sometimes operate on visual feature dimensions, as well as the whole objects. Actually, attention could be allocated to spatial locations, feature dimensions and objects, while we are unable to perfectly control ourselves from attending to specific units.

1.2.3.3   Feature Integration Theory

As discussed above, feature dimension could become a candidate unit of attention. As one of the most influential psychological models of human visual attention, the feature integration theory (Treisman and Gelade, 1980) proposed that various kinds of visual features are registered unconsciously and automatically in the early stage of the perceptual process when perceiving a stimulus. In this theory, the fast "feature search" is performed in parallel as the first stage to pop-out certain feature dimensions. After that, the slow "conjunction search" is performed in serial as the second stage by using the

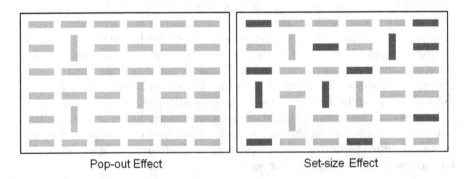

Pop-out Effect   Set-size Effect

Fig. 1.7 The pop-out and set-size effects

combination of the selected features and treat them as a whole for object identification. In these two stages, the latter one requires the involvement of attention while the former one doesn't. Therefore, the "feature search" stage is also called the preattentive stage, while the "conjunction search" stage is called the focused attention stage. The candidate visual features involved in the preattentive stage are named as the preattentive features.

However, it is still unclear what kinds of features can act as the candidates. To address this concern, the probable, possible and unlikely preattentive features are summarized in (Wolfe, 2005). We also list these features in Table 1.1 to facilitate the further reading. In existing researches on visual saliency computation, color and orientation are the most popular spatial features, while flicker and motion are the basic features for temporal saliency analysis.

Table 1.1 Probable, possible and unlikely preattentive features

Probable cases		Possible cases	Unlikely cases
Color	Luminance onset (flicker)	Novelty	Intersection
Orientation	Luminance polarity	Letter identity	Faces
Shape	Size (length & spatial freq.)	Shininess (gloss luster)	Optic flow
Number	Motion (speed/direction)	Alphanumeric category	Color change
Aspect ratio	Stereoscopic depth	Threat	3D volumes
Opacity	Pictorial depth cues		Your name
Curvature	Line termination		
Vernier offset	Lighting direction		
Closure/holes			

1.2.3.4   Contextual Cuing Effect

When computing visual saliency, an important step is to infer the latent top-down modulation mechanisms from training scenes and then transfer the learned "prior knowledge" to testing scenes. Neurobiological studies shows that such mandatory top-down factors may take effect even without the act of subjective will. Thus two important problems arise: why such top-down factors are learnable and how does the knowledge transfer act?

Typically, the visual world is highly structured and scenes are composed of features that are not random. For scenes with similar contents, experimental results show that visual neurons exhibit tuning properties that can be optimized to respond to recurring features in similar visual input (Chun, 2005). In this case, the global visual context guides what to expect and where to look in familiar scenes, allowing subjects to process them more efficiently.

To explain this phenomenon, Fig. 1.8 shows a special example from the surveillance scenario. When watching a surveillance video, people can "learn" implicitly from the viewed scenes about the optimal features and locations to pop-out the vehicles (i.e., what and where), which are surrounded with repeated distractors such as building, lawn and overlaid texts. When encountering similar scenes, people often avoid spending more neural resources in spatial locations and feature dimensions with lower probabilities of containing the targets, leading to a faster response. In psychology, this phenomenon is called the contextual cuing effect, which is defined as an attentional facilitation effect derived from past experiences of the visual world. More importantly, it makes the top-down factors learnable since it support the assumption that similar prior knowledge will be used to process similar scenes.

Suppose that we have learned the prior knowledge on some scenes, yet another problem is how to transfer the knowledge to similar scenes. Generally, it is believed that the invariant properties of the visual environment can be encoded and consulted to guide the knowledge transfer in visual processing. The stable properties, such as rough spatial layout and predicable variation, work as the contextual cues to trig the corresponding prior knowledge that helps processing similar environments.

To facilitate the understanding of the process in our vision system, we summarize a simple flowchart in Fig. 1.9. When information is flowing in the brain, the stable properties of the scene will be encoded in some stages, which usually take the form of global features that can be viewed as a kind of scene labels. These stable or invariant characteristics then enter the working memory and are used to activate the corresponding (procedural or declarative) knowledge stored in the long-term memory through recall and recognition processes. The knowledge stored in the activated neural network will be loaded into the working memory to generate the control signals and launch the mandatory top-down processes, leading to the biased information

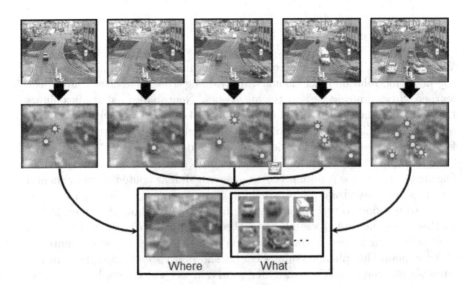

Fig. 1.8 Recurring features serve as contextual cues to process similar scenes

processing. In this process, the top-down factors try to parse and explain the input scenes using the prior knowledge. In particular, this mechanism also acts even when encountering unfamiliar scenes. Actually, the process can be viewed as a learning mechanism of the vision system. On one hand, the past experience can help to efficiently process the familiar scenes. On the other hand, the optimal processing strategy can be mined from unfamiliar scenes to update the knowledge network in the brain.

1.2.4   Summary

From the discussions above, we can summarize that the human vision system has the following characteristics that are tightly correlated with visual saliency computation:

1. **Objective**. Existing studies on computational attention/saliency models mainly focus on the selective characteristic of attention. Consequently, the main objective of visual saliency computation can be described as measuring the importance of various visual subsets by popping-out the targets and inhibiting the distractors.
2. **Processing Mechanisms**. Two processing mechanisms exist in human vision system. The bottom-up mechanism starts earlier and acts faster than the top-down mechanism. The top-down mechanism takes effect from LGN, which is an early stage of the bottom-up information flow. Its influence gradually becomes stronger in the later stages.

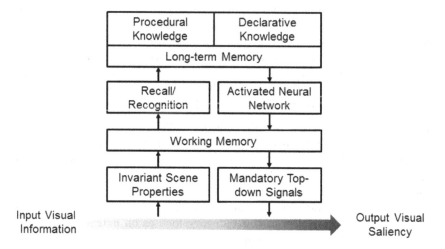

Fig. 1.9 Invariant scene properties are encoded to trig the top-down modulation by activating the prior knowledge learned from similar scenes

3. **Signal Types**: In retina, the input signals can be represented as various preattentive features, which are greatly compressed by extracting and transmitting only the most informative parts. The compressed signals are further processed in LGN to form two categories of signals with different spatial/temporal resolutions, which are fed into two visual pathways for further processing.

4. **Processing Units**. Attention could be allocated to spatial locations, feature dimensions and objects, while we are unable to perfectly control ourselves from attending to specific units. Thus it is reasonable to apply top-down modulations on locations, objects and feature dimensions.

5. **Top-down Factors**. Two top-down factors are involved in the information process, including the volitional one and the mandatory one. The volitional factor is related to active will (e.g., task). The mandatory factor is related to experience and prior knowledge, which can be learned from scenes with similar invariants according to the contextual cuing effect.

6. **Top-down Influences**. Two kinds of top-down influences may exist in the brain, including the additive baseline shifts and multiplicative modulation gains. These two influences may both take effect in computing visual saliency.

1.3 How to Compute Visual Saliency

From the key findings in neurobiology and psychology, we can see that human brain is an extremely complex system and studies on the vision system still stay at the early stage. Several mechanisms can be observed in neurobiological and psychological experiments, while only a small portion of the bottom-up mechanisms have been sufficiently studied and can be directly simulated. These approaches that directly simulate the bottom-up mechanisms of human vision system can be roughly referred to as the bottom-up approaches. In this book, we will describe several representative bottom-up approaches. According to the saliency units used in these approaches, we can roughly group them into two categories, including:

1. Location-based category. In this category, saliency is assumed to be assigned to specific locations (e.g., pixels or macro-blocks). In this book, we will introduce the technical details of several representative approaches to show how to detect salient locations from the spatial domain, the transform domain and the spatiotemporal domain.

2. Object-based category. In this category, saliency is directly assigned to objects. Some approaches in this category focuses on extracting the entire salient objects from the location-based saliency maps, while other approaches aim to directly measure the visual saliency in object-level. This book will introduce several approaches that utilize each of the solutions. In particular, we will introduce two innovative approaches to demonstrate

how to extract salient objects from complementary location-based saliency maps, and how to directly compute the object-based saliency through single image optimization.

One characteristic of the bottom-up approaches is that the cues which are mainly from the input scene will be put into consideration when computing visual saliency. However, the details of most vision mechanisms, especially the top-down components, remain far from being clarified. Thus it is extremely difficult to directly simulate these mechanisms. To address this problem, machine learning algorithms have been introduced and the main objective is to automatically infer and simulate these mechanisms in human vision system. These approaches, which have to learn the cues from other scenes for estimating visual saliency of the input scene, are named as the top-down approaches. Note that the top-down approaches may also contain certain bottom-up processes, making them much more complete in the overall framework[3]. In this book, we propose five novel approaches to address three major problems in learning-based saliency computation, including:

1. How to fuse various bottom-up and top-down factors. A perfect saliency model should jointly consider the influences of bottom-up and top-down processes. Toward this end, we present a probabilistic multi-task learning approach for computing visual saliency by simultaneously integrating the bottom-up and top-down factors. In this approach, the bottom-up and the task-related top-down factors, which are learned from user data, are considered simultaneously in a probabilistic framework. Beyond this supervised learning-based approach, a Bayesian approach is also proposed to learn the prior knowledge from millions of images in an unsupervised manner. Such priors are then used to modulate the bottom-up saliency by recovering the wrongly suppressed targets and inhibiting the improperly popped-out distractors. Extensive experiments show that both approaches can give superior performance compared to existing approaches.

2. How to mine the cluster-specific knowledge. It is obvious that we will use different kinds of prior knowledge to handle different scenarios to determine what is salient. Toward this end, we present a multi-task rank learning approach for visual saliency computation. In this approach, we demonstrate that saliency computation can be best formulated as a ranking problem that aims to distinguish targets from distractors. Consequently, a multi-task learning to rank framework is proposed to infer multiple saliency models that apply to different scene clusters. That is, each model encodes the mechanisms that can best distinguish targets from distractors in a specific cluster of scenes. In the training process, this approach can infer multiple visual saliency models simultaneously. With an appropriate sharing of information across models, the generalization ability of each model can be greatly improved.

---

[3] We will give a more precise definition on bottom-up and top-down models in Chap. 3.

3. How to remove the label ambiguity in user data. When training saliency models, one important problem is that the user data recorded by eye tracking devices or manual labeling activities often contain rich label ambiguities. In the training process, such ambiguities should be removed to improve the performance of learned saliency models. Toward this end, we propose two approaches that also adopt the learning to rank framework. The first approach proposes to embed the ambiguous data into the pairwise learning to rank framework in a cost-sensitive manner. The second approach adopts a multi-instance learning to rank framework to iteratively recover the correct sample labels and optimize the saliency model. These two approaches are proved to be effective in removing label ambiguities in user data. Consequently, the performance of visual saliency models trained with these approaches demonstrate impressive improvement.

1.4   Structure of the Book

The remainder of this book is organized as follows: Chap. 2 introduces the benchmarks and evaluation methodologies for visual saliency computation. Chap. 3 is concerned with the bottom-up visual saliency computation approaches that focus on detecting salient locations and features. In contrast, Chap. 4 addresses the problem of detecting salient objects.

From Chaps. 5-7, we will discuss the learning-based visual saliency estimation. In Chap. 5, the problem of visual saliency computation is re-considered from the perspective of machine learning. A probabilistic learning framework and a Bayesian framework are proposed in this Chapter to demonstrate how to model the top-down factors through supervised and unsupervised learning. In Chap. 6, a learning to rank framework is proposed to model the domain-specific knowledge for visual saliency computation. For the problems that arise in these two learning frameworks, Chap. 7 discusses two feasible solutions for the label ambiguity challenge.

In Chap. 8, we introduce several applications that adopt visual saliency for efficient and effective image/video processing. Finally, we summarize the contributions of our work and give an outlook on possible future work in Chap. 9.

References

Anllo-Vento, L., Luck, S., Hillyard, S.: Spatio-temporal dynamics of attention to color: Evidence from human electrophysiology. Human Brain Mapping 6(4), 216–238 (1998), doi:10.1002/(SICI)1097-0193(1998)6:4(216:AID-HBM3)3.0.CO;2-6
Baluch, F., Itti, L.: Mechanisms of top-down attention. Trends in Neurosciences 34(4), 210–224 (2011), doi:10.1016/j.tins.2011.02.003

Chun, M.M.: Contextual guidance of visual attention. In: Itti, L., Rees, G., Tsotsos, J. (eds.) Neurobiology of Attention, 1st edn., pp. 246–250. Elsevier Press, Amsterdam (2005)

Chun, M.M., Jiang, Y.: Contextual cueing: Implicit learning and memory of visual context guides spatial attention. Cognitive Psychology 36(1), 28–71 (1998)

Cisco, Cisco visual networking index: Forecast and methodology, 2008–2013. Tech. rep., Cisco Systems, Inc. (2009)

Cisco, Cisco visual networking index: Forecast and methodology, 2012–2017. Tech. rep., Cisco Systems, Inc. (2013)

Connor, C.E., Preddie, D.C., Gallant, J.L., Essen, D.C.V.: Spatial attention effects in macaque area v4. Journal of Neuroscience 17(9), 3201–3214 (1997)

Eriksen, B.A., Eriksen, C.W.: Spatial attention effects in macaque area v4. Perception and Psychophysics 16(1), 143–149 (1974)

Itti, L., Rees, G., Tsotsos, J.: Neurobiology of Attention, 1st edn. Elsevier Press, Amsterdam (2005)

James, W.: The Principles of Psychology, 4th edn. 2 vols, 1890. Dover Publications (1950)

Kanwisher, N., Wojciulik, E.: Visual attention: Insights from brain imaging. Nature Reviews Neuroscience 1(2), 91–100 (2000)

Koch, C.: Biophysics of Computation: Information Processing in Single Neurons, 1st edn. Oxford University Press, New York (2004)

Koch, K., McLean, J., Segev, R., Freed, M.A., Berry, M.J., Balasubramanian, V., Sterling, P.: How much the eye tells the brain. Current Biology 16(14), 1428–1434 (2006), doi:10.1016/j.cub.2006.05.056

Le Meur, O., Le Callet, P., Barba, D., Thoreau, D.: A coherent computational approach to model bottom-up visual attention. IEEE Transactions on Pattern Analysis and Machine Intelligence 28(5), 802–817 (2006), doi:10.1109/TPAMI.2006.86

Luria, A.R.: The Working Brain. Basic Books (1973)

O'Craven, K.M., Rosen, B.R., Kwong, K.K., Treisman, A., Savoy, R.L.: Voluntary attention modulates fMRI activity in human MT-MST. Neuron 18, 591–598 (1997)

O'Craven, K.M., Downing, P.E., Kanwisher, N.: fMRI evidence for objects as the units of attentional selection. Nature 401, 584–587 (1999)

Palmer, S.E.: Vision Science: Photons to Phenomenology, 1st edn. MIT Press, Cambridge (1999)

Raichle, M.E.: The brain's dark energy. Scientific American Magazine, 44–49 (2010)

Shepherd, G.M.: The Synaptic Organization of the Brains, 5th edn. Oxford University Press, New York (2003)

Sohlberg, M.M., Mateer, C.A.: Introduction to cognitive rehabilitation: Theory and practice. Guilford Press, New York (1989)

Solso, R.L.: Cognitive Psychology, 2nd edn. Allyn and Bacon, Boston (1988)

Sternberg, R.J., Sternberg, K., Mio, J.: Cognitive Psychology, 6th edn. Cengage Learning (2008)

Treisman, A.M., Gelade, G.: A feature-integration theory of attention. Cognitive Psychology 12(1), 97–136 (1980)

Wilkinson-Berka, J.L.: Diabetes and retinal vascular disorders: role of the renin-angiotensin system. Expert Reviews in Molecular Medicine 6, 1–18 (2004), doi:10.1017/S1462399404008129

Wolfe, J.M.: Guidance of visual search by preattentive information. In: Itti, L., Rees, G., Tsotsos, J. (eds.) Neurobiology of Attention, 1st edn., pp. 101–104. Elsevier Press, Amsterdam (2005)

Wolfe, J.M., Alvarez, G.A., Horowitz, T.S.: Attention is fast but volition is slow. Nature 406, 691 (2000)

Wundt, W.M.: Grundzüge der physiologischen Psychologie, 4th edn., vol. 2. W. Engelman, Leipzig (1893)

Chapter 2
Benchmark and Evaluation Metrics

This Chapter reviews the benchmarks and evaluation metrics for visual saliency computation. It consists of four main parts. The first part provides an overview of saliency model evaluation. In the second part, we review existing saliency benchmarks. Based on these benchmarks, the third part describes various evaluation metrics for measuring the performance of saliency models. The last part concludes this Chapter and proposes how to achieve impressive performance when using these benchmarks and evaluation methodologies.

2.1 Overview

In visual saliency computation, a saliency map is often generated to reveal the salient subsets in an image or video frame. As shown in Fig. 2.1, such saliency maps may be represented in two different forms. In the first form, a saliency map is represented as a grey map with each pixel falling in the dynamic range of $[0, 255]$. A pixel with higher grey value is considered to be more salient. This form of saliency map is usually used for location-based saliency computation. In the second form, a saliency map is represented as a binary mask with 1 for salient pixel and 0 for background pixel. This form of saliency map is usually used for object-based saliency computation.

With the numerous approaches proposed to compute visual saliency, performance evaluation becomes a critical issue. The evaluation objective can be described as finding whether the saliency maps given by a saliency model can accurately reflect the importance of various visual subsets. Generally speaking, an evaluation strategy for visual saliency involves at least two aspects: benchmark and evaluation methodologies. Among them, benchmark can provide us an open testing platform to evaluate new approaches against older ones. Usually, such saliency benchmark can be constructed through subjective evaluation. Given the benchmark, we need a set of appropriate

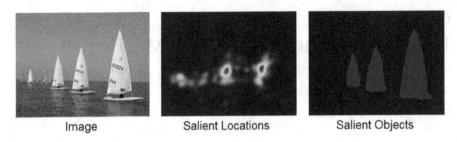

Image   Salient Locations   Salient Objects

Fig. 2.1 Location-based and object-based visual saliency maps

criteria and metrics to quantitatively evaluate the performances of different approaches for visual saliency computation. Therefore, in the next two parts of this Chapter, we will review existing saliency benchmarks and evaluation methodologies, separately.

2.2   Benchmarks

In the field of psychology, there exist numerous image/video datasets that are related with human visual attention or saliency. However, many of them are designed for verifying specific mechanisms of human vision system, such as the gaze policies when playing TV games (Peters and Itti, 2009), making tea (Land et al, 1999) and driving in urban regions (Land, 2006). Yet many of them are only suitable for subjective evaluation, not for quantitative comparison. In this book, we only focus on the public benchmarks obtained in free-viewing conditions for quantitative attention/saliency evaluation. As shown in Table 2.1, these benchmarks can be divided into two main categories, including the image-based category and the video-based category. Most existing benchmarks are image-based since the amount of data to be processed in constructing the image benchmark is much less than that of video. To facilitate the reading, we also generate a "ground-truth" saliency map for each image or video frame by using the user data, which indicates the fixation density map or binary object mask in different benchmarks.

2.2.1   Image Benchmark

When building saliency benchmarks, one of the most straightforward approaches is to record the human fixations when watching images or videos. For example, Einhäuser et al (2006) constructed a benchmark with 108 grey images with resolution of 1024×768. As shown in Fig. 2.2, all these images are

Table 2.1 Benchmark overview

Visual stimulus		User data	Representative Benchmark	Scale
Image Benchmark	Gray image	Fixation	Einhäuser et al (2006)	108 images
	Color image	Fixation	Bruce and Tsotsos (2005)	120 images
	Color image	Fixation	Judd et al (2009)	1,003 images
	Color image	Description	Itti et al (1998)	~1,000 images
	Color image	Rectangle	Liu et al (2007)	5,000 images
	Color image	Mask	Achanta et al (2009)	1,000 images
Video Benchmark	Color video	Fixation	Itti (2008)	25 minutes
	Color video	Rectangle	Li et al (2009)	7.5 hours

about natural scenes. Eight contrast modification operations were then conducted on the original images to generate $108 \times 9 = 972$ trails. For each trail, fixations were recorded from two rhesus monkeys and seven undergraduate students (three male and four female). Note that each human subject only participated in part of the trails while monkeys performed all 972 trails. Due to the limitation of eye tracking devices, only the coordinate of each fixation was recorded. That is, each fixation is linked with only one pixel in the screen without any scale information.

Since the fixations can well reveal the distribution of salient visual subsets (Land, 2006), the ground-truth saliency map can be approximated by the fixation density map. However, the fixation number recorded in a short viewing period is often very limited and subjects actually attend to regions instead of pixels. Therefore, the fixation density map can be generated by summing up a set of 2D Gaussians centered at each fixation (Harel et al, 2007; Wang et al, 2010). Since the fixation density map built with this approach can recover the salient regions from sparse fixations, the authors of existing saliency benchmarks often provide such maps along with the raw fixations, while the parameters of 2D Gaussians are set according to the experimental settings in gathering the user data.

The grey image benchmark has some advantages in evaluating saliency models. However, it also has three drawbacks. First, all images in the dataset are grey, making it difficult to explore the influence of color information. Second, most images in this benchmark have no obvious foreground objects and lack the ability to provide intuitive impression on the performance. Third, many realistic scenarios in computer vision are not limited to natural scenes, which restricts the usage of this benchmark.

To address these three problems, Bruce and Tsotsos (2005) proposed a color image benchmark. This benchmark contains 120 color images from indoor and outdoor scenarios with a resolution of 681×511. As shown in Fig. 2.3, some images contain obvious foreground objects while others have no particular regions of interest. When building the dataset, these 120 images were randomly displayed to several subjects (20 different subjects in total). In each trail, a subject was positioned 0.75m from a 21 inch CRT to watch the image for 4 seconds under the free-viewing conditions. The fixations of

Fig. 2.2 Representative images and ground-truth saliency maps from the benchmark proposed in (Einhäuser et al, 2006)

these subjects were then recorded with non head-mounted eye tracking device, which were then used to build the fixation density maps. Since some images contain very salient objects, this benchmark becomes very popular for both subjective and quantitative comparisons in (Hou and Zhang, 2009; Wang et al, 2010; Gao et al, 2009b).

Fig. 2.3 Representative images and ground-truth saliency maps from the benchmark proposed in (Bruce and Tsotsos, 2005)

One main drawback of this benchmark is that it only contains 120 images, which are far from sufficient for evaluating the learning-based models. When learning the top-down factors on such a small benchmark, severe over-fitting problem may arise. The same problem may also arise to some other image benchmarks proposed in (Kienzle et al, 2007; Cerf et al, 2009) which only have limited number of images. To evaluate the learning-based models, a much larger benchmark is proposed in (Judd et al, 2009) which consists of 1,003 images randomly crawled from Flickr and LabelMe (Russell et al, 2008). Each image has a max side length of 1024 pixels. In the experiments, each image was displayed on a 19 inch screen for 3 seconds to collect the fixations from 15 subjects who free-viewed the image. These fixations were then convolved with a Gaussian filter to generate the fixation density maps. Compared with the benchmark in (Bruce and Tsotsos, 2005), this benchmark is much larger and the scenarios are much more complex, making it more suitable for evaluating the learning-based models.

From the discussions above, we can see that the fixation-based benchmark can perfectly reveal the salient locations. However, the viewing time on each image is often very short (i.e., only 3-4 seconds per image). In such a short time period, the number of collected fixations is very small, making it difficult to reveal the large salient object as a whole. To solve this problem, some researchers discard the eye-tracking device and turn to the manual labeling tools to build the regional saliency benchmarks (Hu et al, 2005; Park and Moon, 2007; Gao et al, 2009a; Huang et al, 2009; Liu et al, 2007; Itti and Koch, 2001; Hou and Zhang, 2007; Itti and Koch, 2000; Itti et al, 1998). In these benchmarks, the most salient object in each image is manually labeled by one or several subjects. For example, Itti et al provided a set of images with obvious salient objects in (Itti et al, 1998; Itti and Koch, 2000, 2001). As shown in Fig. 2.4, these images are mainly indoor/outdoor scenes with text descriptions of the salient objects (e.g., traffic logos, red cans or horizontal bars). Many saliency approaches such as (Guo et al, 2008; Wang et al, 2010; Orabona et al, 2005) adopted these images to demonstrate model performance in an intuitive manner. However, this benchmark cannot be directly used for quantitative comparison.

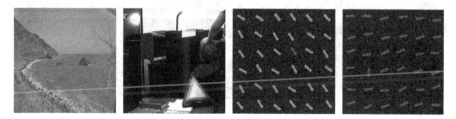

Fig. 2.4 Representative images from the benchmark proposed in (Itti et al, 1998; Itti and Koch, 2000, 2001)

To perform quantitative comparisons, some image benchmarks manually labeled the salient targets with rectangles or accurate object masks. The ground-truth saliency maps are then generated using these labeling results. The most popular benchmark of this type is provided by (Liu et al, 2007) and (Achanta et al, 2009). Liu et al (2007) first collected a 130,099 high quality images and then manually selected 20,840 images for manual labeling.

Each selected image contains a salient or distinctive object, while those with large objects are excluded. In the labeling process, three users were asked to label the most salient object in each image with one rectangle. After that, 5,000 images with highly consistent labeling results were selected and nine different users were asked to label the salient objects again. Each subject was asked to draw only one rectangle for the most salient object in each image. As shown in the first two rows of Fig. 2.5, these images have less ambiguity of what the salient objects are, making the benchmark useful for evaluating the performance of object-based visual saliency computation.

However, one drawback of this benchmark is that it lacks the accurate object mask and sometimes multiple objects are clubbed into a big rectangle. To address this problem, Achanta et al (2009) further selected 1,000 images from these 5,000 images with highly consistent salient objects. Salient objects in these 1,000 images are manually segmented to obtain binary object masks. As shown in the third row of Fig. 2.5, these masks can reveal the accurate boundaries of salient objects, which performs better than rectangles in evaluating object-based saliency model.

Fig. 2.5 Representative images and ground-truth saliency maps from the benchmark proposed in (Liu et al, 2007; Achanta et al, 2009)

Beyond the benchmarks discussed above, there are some other image benchmarks. For example, Li et al (2013a) proposed a benchmark with 265 images. These images are grouped into six categories, while each category has different difficulties for saliency computation. For each image, they record the fixations from 21 subjects, while the accurate object masks are also provided. Ramanathan et al (2010) provided a benchmark with 758 images containing semantically affective objects or scenes and fixations are recorded from 75 subjects. Huang et al (2009) presented a collaborative benchmark with 993 images. By collecting and averaging a large number of rectangular annotations from an interactive game, multiple salient objects can be robustly labeled in each image. Cerf et al (2009) collected fixation data from 8 subjects doing a free-viewing task on 180 color outdoor and indoor images. These images include many different types of faces and observers were also asked to rate how interesting each image was. There are many other image benchmarks and due to the limitation of the book length, we cannot discuss them one by one. Readers can refer to (Borji et al, 2012) to get more details.

2.2.2 Video Benchmark

Beyond image benchmarks, some approaches (Itti, 2008; Itti and Baldi, 2005b; Peters and Itti, 2007; Marat et al, 2009) have recorded fixations on video to verify the influence of temporal information. Among these benchmarks, the most popular two that are publicly available can be denoted as **ORIG** and **MTV**. The **ORIG** benchmark is proposed in (Itti, 2004), which has been used by many spatiotemporal saliency models for performance evaluation (Itti, 2005; Itti and Baldi, 2005a; Wang et al, 2010). The **MTV** benchmark is built on the basis of the **ORIG** dataset, and is used in the psychological studies (Carmi and Itti, 2006a,b). In the year of 2008, these two benchmarks were published online for free downloading (Itti, 2008).

The **ORIG** benchmark consists of over 46,000 video frames in 50 video clips (25 minutes in total). These videos contain scenes such as "outdoors day & night," "crowds," "TV news," "sports," "commercials," "video games" and "test stimuli." The eye traces of 8 subjects (5 male and 3 female, aged 23-32) when watching these clips (4 to 6 subjects/clip) were recorded using a 240HZ ISCAN RK-464 eye-tracker. Some representative video frames and fixation density maps are shown in Fig. 2.6, from which we can see that the fixations in each video frame are very sparse.

Compared with the **ORIG** benchmark, the **MTV** benchmark is constructed using the same video data but different approaches. To distinguish the influences of bottom-up and top-down factors in human vision system, the videos are selected from the **ORIG** benchmarks and cut into 1-3s clippets. These clippets are then randomly reassembled to form new videos (Carmi and Itti, 2006a,b). Fixations on these videos were recorded using the same approach as in building the **ORIG** benchmark, but with different subjects. In conclusion, a video in the **ORIG** benchmark has consistent scenario, while a video from the **MTV** benchmark consists of multiple scenarios that are switching rapidly. Therefore, the **MTV** benchmark is more suitable for studying the influence of temporal information.

Generally speaking, video benchmarks contain more frames than the image benchmarks and are much suitable for learning-based approaches. However,

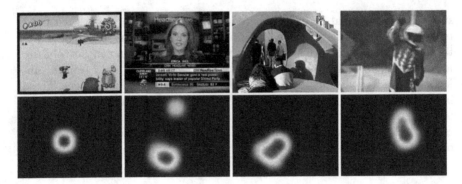

Fig. 2.6 Representative video frames and ground-truth saliency maps from the benchmark proposed in (Itti, 2008)

we can see that the fixations collected from each video frame are much less than those gathered from an image. Recall that an image will display for 4 seconds in (Bruce and Tsotsos, 2005) to collect fixations, while video frames are usually displayed at around 30 frames per second. In such a short time period, the recorded fixations can only reveal part of the most salient object (as shown in Fig. 2.6). Therefore, it is necessary to construct a video benchmark that can reveal the whole salient object in each frame. Toward this end, Li et al (2009) proposed a regional saliency benchmark for video. In this benchmark, videos are collected from four main categories:

1. **Surveillance video**. In most cases, surveillance video contains static backgrounds and dynamic salient objects, which can be used for visual saliency analysis. In our dataset, surveillance videos are selected from the CAVIAR dataset (CAVIAR, 2004).
2. **Artificial video**. To explore the differences of visual saliency in natural and artificial scenes, artificial video clips are selected from 2D and 3D cartoons.
3. **Natural video with artificial parts**. Usually, the artificial parts in natural video such as captions and logos have a strong impact on visual saliency. Videos with artificial parts are collected from TRECVID 2006/2007 and the Internet.
4. **Natural video**. Similarly, natural videos with no artificial parts such as overlaid captions and logos are selected from TRECVID 2006/2007 and the Internet.

In total, the video benchmark contains 431 short videos with a total length of 7.5 hours, in which 764,806 frames are involved. The benchmark mainly covers videos from six genres: documentary, ad, cartoon, news, movie and surveillance. Among these videos, salient objects in surveillance videos have been labeled in (CAVIAR, 2004). For other videos, 23 subjects (17 male and 6 female, aged 21-37, 10-23 subjects for each clip) were asked to label the salient object(s) in video. Since it is invincible to manually label all frames, only 62,356 key frames were selected from these videos for labeling (I frames for MPEG videos or sampling one frame out of every 15 frames for other videos). In the labeling process, subjects were first instructed to free-view a short video. Then the key frames of the former video were displayed again in chronological order. Subjects were asked to label all regions that they thought to be salient in previous watching with one or multiple rectangles. After labeling, the ground-truth saliency maps can be obtained by combining the labeling results of all subjects, which were further smoothed with a 2D Gaussian kernel. Some representative examples can be found in Fig. 2.7. We can see that this labeling method can reveal multiple salient objects simultaneously, making it suitable for evaluating object-based saliency models in the spatiotemporal domain.

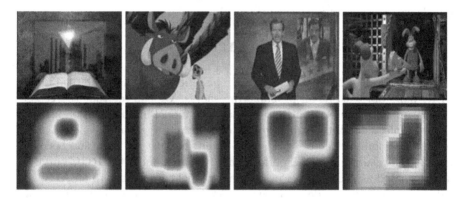

Fig. 2.7 Representative video frames and ground-truth saliency maps from the benchmark proposed in (Li et al, 2009)

2.3   Evaluation Metrics

From the discussions above, we can see that there already exist many image and video benchmarks for video saliency evaluation. Thus another problem is, what kinds of evaluation methodologies should be used for quantitative comparison? In this subsection, we will introduce these evaluation metrics according to the popularity, and then discuss which methodologies should be used for evaluation.

2.3.1   Receiver Operating Characteristic

The Receiver operating characteristic (**ROC**) is the most popular evaluation metric to measure the performance of visual saliency models. In the evaluation, **ROC** takes a saliency model as a binary classifier and evaluate its performance under different criteria (Shic and Scassellati, 2007; Kienzle et al, 2007; Hou and Zhang, 2009). Suppose that the predicted saliency maps are normalized to the dynamic range of [0,255], the evaluation will be conducted by using all probable thresholds in $\{0, 1, ..., 255\}$. On each threshold, the predicted saliency maps are binarized into foreground and background regions to calculate the number of True Positives (TP), True Negatives (TN), False Positives (FP) and False Negatives (FN):

$$\text{TP} = \#(foreground \ \& \ fixated), \ \text{FP} = \#(foreground \ \& \ non\text{-}fixated),$$
$$\text{FN} = \#(background \ \& \ fixated), \ \text{TN} = \#(background \ \& \ non\text{-}fixated).$$
$$(2.1)$$

Consequently, the True Positive Rate (TPR) and False Positive Rate (FPR) can be calculated at each threshold:

$$\text{FPR} = \frac{\text{FP}}{\text{FP} + \text{TN}}, \ \text{TPR} = \frac{\text{TP}}{\text{TP} + \text{FN}}. \tag{2.2}$$

As shown in Fig. 2.8, a **ROC** curve can be generated by using all the (TPR, FPR) pairs obtained at all probable thresholds. We can see that the **ROC** curve can reveal the performance of a saliency model in different conditions. For example, a small threshold will pop-out more targets and lead to higher recall. But more distractors will also pop-out, making the precision become lower. To measure the overall performance, the area under the **ROC** curve is used, denoted as **AUC**. A perfect saliency model corresponds to an **AUC** of 1.0, while a random model will have an **AUC** of 0.5. For more details about **ROC**, please refer to (Fawcett, 2006).

When using **ROC** and **AUC**, there are usually two ways to evaluate the overall performance on multiple images: 1) calculate the **AUC** score on each image first and then compute the mean and standard deviation of all the **AUC** scores; and 2) sum up the numbers of true positives, true negatives, false positives and false negatives on all images in the benchmark and generate a unique **ROC** curve, leading to a unique **AUC** score. Both ways can make sense and the first way is preferred in recent studies.

In the evaluation, different saliency models often generate saliency maps with different resolutions. Therefore, there are two feasible ways to perform the comparison. The first way is to re-scale the recorded fixations to the resolutions of saliency maps and conduct the evaluation in the macro-block level. The evaluation performed on the block-level can be extremely efficient, but the **AUC** score may be somewhat higher than normal for the algorithms that use larger blocks. The second way is to resize all the saliency maps to the original resolutions of the input images for fair comparison. Thus each pixel can be classified as "fixated" or "non-fixated" using its saliency value, which can be verified by the recorded fixations. Evaluation in this way is much more accurate and fairer than the block-based evaluation.

When computing **AUC**, the central fixation bias is an important issue. That is, human fixations are often biased to image centers while non-fixated pixels usually distribute around image edges. However, the different distributions of fixated and non-fixated pixels often lead to unfair comparisons by favoring the saliency models that mainly highlight the targets around image centers (e.g., using center-bias re-weighting) or ignore distractors near to image borders (e.g., cutting the borders of the predicted saliency maps). For instance, Judd et al randomly selected 10 fixated and 10 non-fixated pixels from the top 20% and the bottom 70% salient pixels on 100 images of the benchmarks from (Judd et al, 2009). They further divided each image into center region and peripheral region, while the center region lies in a circle around image center whose radius equals to 42% of the distance from image center to image corner. After the division, the center region contains 78.8% fixated pixels and 24.5% non-fixated pixels, while the numbers change to 21.2% and 75.5% in the peripheral region, respectively. In this case, a model that simply emphasizes the center region will pop-out most of the fixated pixels and suppress most of the non-fixated pixels, leading to unfair comparisons.

Inspired by the approach used in (Tatler et al, 2005), a feasible solution to this problem can be randomly re-sampling the non-fixated pixels according to

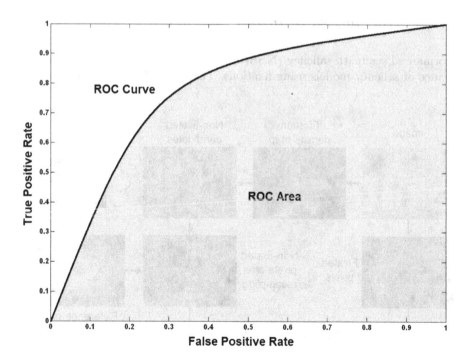

Fig. 2.8 Example of receiver operating characteristic (ROC) curve

the distribution of fixations on all the images in the same benchmark. In the re-sampling process, as shown in Fig. 2.9, non-fixated pixels are re-sampled from those with saliency scores lower than a threshold (e.g., 5% of the maximum saliency). In this manner, possible ambiguities, such as simultaneously selecting fixated and non-fixated pixels from the same object, can be largely avoided. For these candidate pixels, a reference map is generated by summing up all the fixation density maps from all the images in the same benchmark to guide the re-sampling process. Note that different benchmarks may have different reference maps due to different experimental settings (e.g., viewing distance, angle and image/screen resolution). A non-fixated pixel will be selected with a high probability if the corresponding pixel in the reference map has a high score. Finally, only the selected non-fixated pixels, which are also biased to image centers, will be used for performance evaluation. After performing the re-sampling operations on the same 100 images used in Judd's experiment, the center region contains 71.0% fixated pixels and 64.1% non-fixated pixels, while the numbers change to 29.0% and 35.9% in the peripheral region, respectively. When the ratios of fixated and non-fixated pixels are comparable in each region, emphasizing only the center region will no-longer obtain much gain, making the comparisons much fairer.

2.3.2 Normalized Scanpath Saliency

Normalized scanpath saliency (**NSS**) is another way to measure the performance of saliency models using fixations. The basic idea is normalizing the

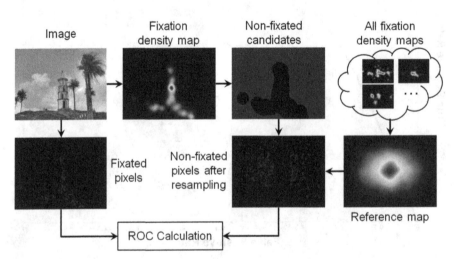

Fig. 2.9 Re-sampling non-fixated pixels for fair ROC evaluation. Reprinted from (Li et al, 2013b), with kind permission from Springer Science+Business Media.

saliency scores of all regions in an image to have zero mean and unit standard deviation. Then the saliency scores at fixated locations are used to measure the model performance (Pang et al, 2008; Peters and Itti, 2007, 2008, 2009). We can see that **NSS** is a kind of z-score, which expresses the divergence of the experimental result from the mean as a number of standard deviations. The larger the value, the less probable experimental result is due to chance. **NSS** can be calculated as:

$$\mu = \frac{1}{N} \sum_{n=1}^{N} s_n,$$

$$\sigma = \sqrt{\frac{1}{N-1} \sum_{n=1}^{N} (s_n - \mu)^2}, \tag{2.3}$$

$$\mathbf{NSS} = \frac{s_e - \mu}{\sigma},$$

where s_n and s_e are saliency scores at the nth location and the fixated location, respectively. N is the total number of candidate locations in the scene. We can see that **NSS**=1 means that the subjects' fixations fall in a region whose predicted saliency is one standard deviation above average.

To simplify the calculation of **NSS** when there exist multiple fixations on one frame, Marat et al (2009) proposed another approach to compute **NSS**. In their approach, the mean and standard deviation are computed as in (2.3), while the computation of **NSS** is simplified as:

$$\mathbf{NSS} = \frac{\frac{1}{N} \sum_{n=1}^{N} g_n \times s_n - \mu}{\sigma}, \tag{2.4}$$

where g_n is the value at the nth location of the fixation density maps. Note that the fixation density map here is also normalized to have unit mean. From (2.3) and (2.4), we can see that a good saliency model has positive **NSS** (the larger the better) and exhibits significantly higher saliency values at fixated locations compared to other locations. Meanwhile, **NSS**=0 indicates that the model performs no better than random model and predicting human fixations by chance. Negative **NSS** means fixation tends to be on non-salient locations predicted by the model.

2.3.3   Kullback-Leibler Divergence

Kullback-Leibler divergence (**KLD**) is an entropy-based evaluation criteria which is often used to evaluate the consistency between predicted saliency maps and gaze paths (Itti and Baldi, 2005a,b; Peters and Itti, 2008). This evaluation metric is motivated by comparing the saliency distributions in fixated points and random points. A good saliency model should assign high saliency values to fixated regions while inhibiting the responses from background regions. Thus the difference between the saliency histograms from fixated and random regions should be large. To quantize such difference, the predicted saliency values are sampled from the fixated points and random points. These values are then normalized to $[0,1]$ and quantized to q bins to form two histograms. The histogram H represents the saliency distribution around fixation points, while histogram R reveals the saliency distribution near random points. Consequently, the **KLD** can be calculated as:

$$\mathbf{KLD} = \frac{1}{2} \sum_{i=1}^{q} \left(H_i \log \frac{H_i}{R_i} + R_i \log \frac{R_i}{H_i} \right), \tag{2.5}$$

where H_i and R_i are the ith component for the saliency histograms in fixated and random regions, respectively. From this definition, we can see that the **KLD** score computed here is symmetric and a larger **KLD** implies a saliency map with compact foreground and clean background.

As proposed in (Borji and Itti, 2013), one main advantage of **KLD** is that it is invariant to reparameterization such as applying continuous monotonic nonlinearity (e.g., $s \leftarrow s^3$) to the predicted saliency. However, one disadvantage of **KLD** is that it lacks a well-defined upper-bound. Two histograms that are completely non-overlapping will lead to a **KLD** score of infinity.

2.3.4   Recall and Precision

All the three evaluation methodologies proposed above focus on evaluating the location-based saliency models using human fixations. However, when using benchmarks with salient rectangles or accurate object masks, we also have to measure whether the salient objects can be detected as a whole. Toward this end, the metrics **Recall** and **Precision** are proposed to quantize the performance of saliency models in object level (Gao et al, 2009a; Liu et al, 2007; Gopalakrishnan et al, 2009; Liu et al, 2008; Chalmond et al, 2006). In the evaluation, the predicted saliency map is often resized to have the same resolution of the ground-truth saliency map for pixel-wise comparison. Note that here the predicted saliency falls in the dynamic range of $[0,255]$, while the ground-truth saliency is binary where $g_n=1$ indicates that the nth pixel is salient.

In the evaluation, there are two ways to calculate the **Recall** and **Precision**. In the first way, all probable thresholds in $\{0, 1, ..., 255\}$ are used to generate a set of "**Recall-Precision**" pairs. When using threshold T_o, the **Recall** and **Precision** can be calculated as:

$$\textbf{Recall} = \frac{\sum_{n=1}^{N} [s_n > T_o]_{\mathbf{I}} \times g_n}{\sum_{n=1}^{N} g_n},$$

$$\textbf{Precision} = \frac{\sum_{n=1}^{N} [s_n > T_o]_{\mathbf{I}} \times g_n}{\sum_{n=1}^{N} [s_n > T_o]_{\mathbf{I}}},$$

(2.6)

where $[\mathbf{e}]_{\mathbf{I}} = 1$ if event $\mathbf{e}$ holds, otherwise $[\mathbf{e}]_{\mathbf{I}} = 0$. Given the model performance when using various thresholds, a "**Recall-Precision**" curve can be generated for evaluating the overall performance.

In the second way, an adaptive threshold is derived on each image to binarize the predicted saliency maps. In (Achanta et al, 2009; Perazzi et al, 2012; Cheng et al, 2011), such intelligent threshold T_I is calculated as:

$$T_I = \frac{2}{N} \sum_{n=1}^{N} s_n.$$

(2.7)

With the adaptive threshold, a unique **Recall** and **Precision** can be calculated on each benchmark. To further measure the overall performance with one unique score, the $\mathbf{F}_\beta$ can be calculated as:

$$\mathbf{F}_\beta = \frac{(1 + \beta^2) \times \textbf{Precision} \times \textbf{Recall}}{\beta^2 \times \textbf{Precision} + \textbf{Recall}},$$

(2.8)

where β is a parameter to balance the influence of **Recall** and **Precision**. Achanta et al (2009), Cheng et al (2011) and Perazzi et al (2012) argued that **Precision** is more important than **Recall** and thus set $\beta^2 = 0.3$. However, we find that **Recall** is as important as **Precision** and β should be set to 1 to equally emphasize these two evaluation metrics. In an experiment, we find that both the location-based and object-based saliency models can easily give promising **Precision**. However, a high **Recall** could be reached only by conducting the object segmentation operations either before or after the saliency computation. Without segmenting images into objects, the **Recall** of a location-based saliency model will be very poor, even when the **Precision** is still promising. We will further discuss this issue in Chap. 4.

2.3.5 Other Metrics

Beyond the aforementioned evaluation metrics, there are many other metrics for the quantitative evaluation, including:

•**Linear correlation coefficient**. The linear correlation coefficient (**CC**) is measured by directly comparing the predicted saliency map S and ground-truth saliency map G. As proposed in (Borji et al, 2013), such relationship can be measured as:

$$\mathbf{CC} = \frac{cov(S, G)}{std(S) \times std(G)},\tag{2.9}$$

where $cov(\cdot)$ and $std(\cdot)$ represent the calculation of covariance and standard deviation, respectively. We can see that there may exist a perfect linear relationship when $\mathbf{CC}$ approximates to $+1/-1$.

•**Percentile**. The metric **Percentile** is proposed in (Peters and Itti, 2008) which is motivated by measuring the percentage of fixations whose predicted saliency values fall below the value at a fixated location. The **Percentile** can be calculated as:

$$\mathbf{Percentile} = 100 \times \frac{|\{s_n < s_e, n = 1, \ldots, N\}|}{N},\tag{2.10}$$

where s_n and s_e are the saliency scores at the nth location and the fixated location, respectively. N is the total number of candidate locations. Note that the **Percentile** is computed on each fixation (or saccade's end point) separately and the overall performance is measured by the mean and standard deviation. Perfect model corresponds to a score of 100%, while random predictions generate a score of 50%.

•**Boundary displacement error**. This evaluation metric can be used if the salient object is labeled with rectangle. When using this metric, the saliency model should further output a rectangle that covers the salient object in the predicted saliency map (Liu et al, 2007, 2008). Then the boundary displacement error (**BDE**) is computed by measuring the average displacement of corresponding boundaries of two rectangles. For the benchmark that receives multiple rectangles from more than one subject on a single image, the **BDE** can be calculated and averaged over different subjects. Note that a fundamental assumption here is a subject will label the salient object in each image with only one rectangle.

•**Subjective evaluation**. When there is no benchmark for a specific scenario, subjective scoring is often used for model comparison (Yu and Wong, 2007; Han et al, 2004; Cheng et al, 2005; Rapantzikos et al, 2007; Li and Lee, 2007; Zhai and Shah, 2006). In the comparison, two different scoring mechanisms can be adopted. In the first mechanism, two saliency maps are randomly selected and displayed simultaneously in the screen. Subjects should determine which saliency map is better and the best model should outperform all the other models in most pair-wise comparisons. In the second mechanism, subjects are asked to give each saliency map an integer score and model performance can be measured by averaging the scores on the whole benchmark. We can see that the first mechanism is more reasonable since it is somehow difficult for subjects to perform direct scoring in some cases. However, the first mechanism need $O(K^2)$ pair-wise comparisons when there are K saliency models, making the evaluation process really time-consuming.

•**Fixation hit rate**. The motivation of this metric is to select a set of probable fixations each the saliency map (e.g., the top several local maximum) and see how can they hit the salient objects. For example, Itti and Koch (2001) computed the average number of false detections before the predicted fixations reach the target(s). Guo et al (2008) counted the number of salient objects detected in the first five fixations.

•**Indirect evaluation.** In some saliency-based applications such as object recognition (Parikh et al, 2008), the performance of saliency models can be indirectly evaluated by measuring the improvement on recognition performance with or without the saliency models.

2.3.6 Summary

To sum up, there exist many evaluation methodologies for measuring the performance of visual saliency models. These metrics can measure the performance of a saliency model from different aspects. Thus they should be combined for fair evaluation, especially for the location-based models. The main reason is due to the center-bias effect. From the perspective of photographers, salient targets are often placed around the view centers when taking photo. This phenomenon can be found in most of the Internet images. From the perspective of photo viewers, they also tend to start from image centers for searching the targets. In particular, viewers are sometimes requested to fixate on the screen center to start a psychological experiment. Therefore, it is obvious that human fixations or salient objects will have higher probabilities to appear around image centers than borders. This bias is very strong and models can easily achieve high scores by favoring the image centers with center-bias re-weighting or ignoring image borders with the border cut.

To leverage such bias, a revised **ROC** metric has been proposed above by re-sampling the non-fixated locations in a center-biased manner. In this way, both fixated and non-fixated pixels will distribute around image centers, leading to fairer comparisons. However, one drawback of **ROC** is that it only focuses on the saliency ordering, while the saliency amplitude is not taken into account. Zhao and Koch (2011) ever reported that two saliency maps, one with almost clean background and one with noisy background, achieved comparable **AUC** scores. In this case, the noise in the background region, which will give rise to more false alarms, will not affect the saliency ordering. Therefore, it is necessary to bring in more metrics such as **NSS** and **CC** that focus on the saliency amplitude.

2.4 Notes

In this Chapter, we have introduced several benchmarks and evaluation methodologies for visual saliency computation. Most of the benchmarks are constructed for evaluating saliency models under free-viewing conditions. By inspecting the classic evaluation methodologies used on these benchmarks, we can derive several key characteristics that a good saliency model should demonstrate. From the receiver operating characteristic, we can see that a good saliency model should rank the targets higher than distractors. This is compatible with human vision system since an important role of visual saliency is to feed the parallel information into serial processing. In this process, these ranks can serve as a good indicator to determine which visual subsets should be processed with the highest priority. From the normalized scanpath saliency and Kullback-Leibler divergence, we can see that the saliency amplitude is as important as saliency ordering. That is, when the orders of various visual subsets stay unchanged, the gap between the responses at targets and distractors should be as large as possible. In this manner, salient targets can pop-out from distractors with less effort.

When building saliency models, there are two feasible solutions to achieve outstanding performance when using these benchmarks and evaluation methodologies. The first solutions is to manually fine-tune the parameters and conduct biased post-processing (e.g., center-bias re-weighting and border cut) by observing subjects' bias in testing data. This solution is often used in the bottom-up approaches. The second solution is to incorporate the learning algorithms. Instead of fine-tuning the model in an ad-hoc manner, the evaluation methodologies can be directly embedded into the optimization objective. By training the model on part of the benchmark, the bias of such benchmark as well as the latent mechanism adopted by subjects in predicting visual saliency can be automatically derived, leading to impressive performance on the rest data of the benchmark.

References

Achanta, R., Hemami, S., Estrada, F., Süsstrunk, S.: Frequency-tuned salient region detection. In: Preceedings of the IEEE Conference on Computer Vision and Pattern Recognition (CVPR), pp. 1597–1604 (2009), doi:10.1109/CVPR.2009.5206596

Borji, A., Itti, L.: State-of-the-art in visual attention modeling. IEEE Transactions on Pattern Analysis and Machine Intelligence 35(1), 185–207 (2013), doi:10.1109/TPAMI.2012.89

Borji, A., Sihite, D.N., Itti, L.: Salient object detection: A benchmark. In: Proceedings of the 12th European Conference on Computer Vision, pp. 414–429 (2012), doi:10.1007/978-3-642-33709-3_30

Borji, A., Sihite, D., Itti, L.: Quantitative analysis of human-model agreement in visual saliency modeling: A comparative study. IEEE Transactions on Image Processing 22(1), 55–69 (2013), doi:10.1109/TIP.2012.2210727

Bruce, N.D., Tsotsos, J.K.: Saliency based on information maximization. In: Advances in Neural Information Processing Systems (NIPS), Vancouver, BC, Canada, pp. 155–162 (2005)

Carmi, R., Itti, L.: The role of memory in guiding attention during natural vision. Journal of Vision 6(9), 898–914 (2006a), doi:10.1167/6.9.4

Carmi, R., Itti, L.: Visual causes versus correlates of attentional selection in dynamic scenes. Vision Research 46(26), 4333–4345 (2006b), doi:10.1016/j.visres.2006.08.019

CAVIAR, Ec funded caviar project/ist, 37540 (2004), http://homepages.inf.ed.ac.uk/rbf/caviar/

Cerf, M., Harel, J., Einhauser, W., Koch, C.: Predicting human gaze using low-level saliency combined with face detection. In: Advances in Neural Information Processing Systems (NIPS), Vancouver, BC, Canada (2009)

Chalmond, B., Francesconi, B., Herbin, S.: Using hidden scale for salient object detection. IEEE Transactions on Image Processing 15(9), 2644–2656 (2006), doi:10.1109/TIP.2006.877380

Cheng, M.M., Zhang, G.X., Mitra, N., Huang, X., Hu, S.M.: Global contrast based salient region detection. In: Preeedings of the IEEE Conference on Computer Vision and Pattern Recognition (CVPR), pp. 409–416 (2011), doi:10.1109/CVPR.2011.5995344

Cheng, W.H., Chu, W.T., Kuo, J.H., Wu, J.L.: Automatic video region-of-interest determination based on user attention model. In: Preeedings of IEEE International Symposium on Circuits and Systems (ISCAS), pp. 3219–3222 (2005), doi:10.1109/ISCAS.2005.1465313

Einhäuser, W., Kruse, W., Hoffmann, K.P., König, P.: Differences of monkey and human overt attention under natural conditions. Vision Research 46(8-9), 1194–1209 (2006), doi:10.1016/j.visres.2005.08.032

Fawcett, T.: An introduction to roc analysis. Pattern Recognition Letter 27(8), 861–874 (2006), doi:10.1016/j.patrec.2005.10.010

Gao, D., Han, S., Vasconcelos, N.: Discriminant saliency, the detection of suspicious coincidences, and applications to visual recognition. IEEE Transactions on Pattern Analysis and Machine Intelligence 31(6), 989–1005 (2009a), doi:10.1109/TPAMI.2009.27

Gao, D., Mahadevan, V., Vasconcelos, N.: The discriminant center-surround hypothesis for bottom-up saliency. In: Advances in Neural Information Processing Systems, NIPS (2009b)

Gopalakrishnan, V., Hu, Y., Rajan, D.: Random walks on graphs to model saliency in images. In: Preeedings of the IEEE Conference on Computer Vision and Pattern Recognition (CVPR), pp. 1698–1705 (2009), doi:10.1109/CVPR.2009.5206767

Guo, C., Ma, Q., Zhang, L.: Spatio-temporal saliency detection using phase spectrum of quaternion fourier transform. In: Preeedings of the IEEE Conference on Computer Vision and Pattern Recognition (CVPR), pp. 1–8 (2008), doi:10.1109/CVPR.2008.4587715

Han, J., Ngan, K., Li, M., Zhang, H.: Towards unsupervised attention object extraction by integrating visual attention and object growing. In: Preeedings of the IEEE Conference on Image Processing (ICIP), vol. 2, pp. 941–944 (2004), doi:10.1109/ICIP.2004.1419455

Harel, J., Koch, C., Perona, P.: Graph-based visual saliency. In: Advances in Neural Information Processing Systems (NIPS), pp. 545–552 (2007)

Hou, X., Zhang, L.: Saliency detection: A spectral residual approach. In: Preeedings of the IEEE Conference on Computer Vision and Pattern Recognition (CVPR), pp. 1–8 (2007), doi:10.1109/CVPR.2007.383267

Hou, X., Zhang, L.: Dynamic visual attention: Searching for coding length increments. In: Advances in Neural Information Processing Systems (NIPS), pp. 681–688 (2009)

Hu, Y., Rajan, D., Chia, L.T.: Adaptive local context suppression of multiple cues for salient visual attention detection. In: Preeedings of the IEEE International Conference on Multimedia and Expo, ICME (2005), doi:10.1109/ICME.2005.1521431

Huang, T.H., Cheng, K.Y., Chuang, Y.Y.: A collaborative benchmark for region of interest detection algorithms. In: Preceedings of the IEEE Conference
on Computer Vision and Pattern Recognition (CVPR), pp. 296–303 (2009),
doi:10.1109/CVPR.2009.5206765

Itti, L.: Automatic foveation for video compression using a neurobiological model
of visual attention. IEEE Transactions on Image Processing 13(10), 1304–1318
(2004), doi:10.1109/TIP.2004.834657

Itti, L.: Quantifying the contribution of low-level saliency to human eye movements
in dynamic scenes. Visual Cognition 12(6), 1093–1123 (2005)

Itti, L.: Crcns data sharing: Eye movements during free-viewing of natural videos.
In: Collaborative Research in Computational Neuroscience Annual Meeting, Los
Angeles, California (2008)

Itti, L., Baldi, P.: Bayesian surprise attracts human attention. In: Advances in Neural
Information Processing Systems (NIPS), pp. 547–554 (2005a)

Itti, L., Baldi, P.: A principled approach to detecting surprising events in video. In:
Preceedings of the IEEE Conference on Computer Vision and Pattern Recognition
(CVPR), vol. 1, pp. 631–637 (2005b), doi:10.1109/CVPR.2005.40

Itti, L., Koch, C.: A saliency-based search mechanism for overt and covert shifts of
visual attention. Vision Research 40(10-12), 1489–1506 (2000), doi:10.1016/S0042-
6989(99)00163-7

Itti, L., Koch, C.: Feature combination strategies for saliency-based visual attention
systems. Journal of Electronic Imaging 10(1), 161–169 (2001)

Itti, L., Koch, C., Niebur, E.: A model of saliency-based visual attention for rapid
scene analysis. IEEE Transactions on Pattern Analysis and Machine Intelligence 20(11), 1254–1259 (1998), doi:10.1109/34.730558

Judd, T., Ehinger, K., Durand, F., Torralba, A.: Learning to predict where humans
look. In: Preceedings of the IEEE International Conference on Computer Vision
(ICCV), pp. 2106–2113 (2009), doi:10.1109/ICCV.2009.5459462

Kienzle, W., Wichmann, F.A., Scholkopf, B., Franz, M.O.: A nonparametric approach
to bottom-up visual saliency. In: Advances in Neural Information Processing Systems (NIPS), pp. 689–696 (2007)

Land, M., Mennie, N., Rusted, J.: The roles of vision and eye movements in the
control of activities of daily living. Perception 28(11), 1311–1328 (1999)

Land, M.F.: Eye movements and the control of actions in everyday life. Progress in
Retinal and Eye Research 25, 296–324 (2006)

Li, J., Tian, Y., Huang, T., Gao, W.: A dataset and evaluation methodology for visual
saliency in video. In: Preceedings of the IEEE International Conference on Multimedia and Expo (ICME), pp. 442–445 (2009), doi:10.1109/ICME.2009.5202529

Li, J., Levine, M., An, X., Xu, X., He, H.: Visual saliency based on scale-space analysis
in the frequency domain. IEEE Transactions on Pattern Analysis and Machine
Intelligence 35(4), 996–1010 (2013a), doi:10.1109/TPAMI.2012.147

Li, J., Tian, Y., Huang, T.: Visual saliency with statistical priors. International Journal of Computer Vision, 1–15 (2013), doi:10.1007/s11263-013-0678-0

Li, S., Lee, M.C.: Efficient spatiotemporal-attention-driven shot matching.
In: Proceedings of the 15th Annual ACM International Conference on
Multimedia, MULTIMEDIA 2007, pp. 178–187. ACM, New York (2007),
doi:10.1145/1291233.1291275

Liu, T., Sun, J., Zheng, N.N., Tang, X., Shum, H.Y.: Learning to detect a salient
object. In: Preceedings of the IEEE Conference on Computer Vision and Pattern
Recognition (CVPR), pp. 1–8 (2007), doi:10.1109/CVPR.2007.383047

Liu, T., Zheng, N., Ding, W., Yuan, Z.: Video attention: Learning to detect a salient
object sequence. In: Preceedings of the 19th IEEE Conference on Pattern Recognition (ICPR), pp. 1–4 (2008), doi:10.1109/ICPR.2008.4761406

Marat, S., Ho Phuoc, T., Granjon, L., Guyader, N., Pellerin, D., Guérin-Dugué, A.: Modelling spatio-temporal saliency to predict gaze direction for short videos. International Journal of Computer Vision 82(3), 231–243 (2009), doi:10.1007/s11263-009-0215-3

Orabona, F., Metta, G., Sandini, G.: Object-based visual attention: a model for a behaving robot. In: Proceedings of the IEEE Conference on Computer Vision and Pattern Recognition (CVPR) - Workshops, pp. 89–89 (2005), doi:10.1109/CVPR.2005.502

Pang, D., Kimura, A., Takeuchi, T., Yamato, J., Kashino, K.: A stochastic model of selective visual attention with a dynamic bayesian network. In: Preceedings of the IEEE International Conference on Multimedia and Expo (ICME), pp. 1073–1076 (2008), doi:10.1109/ICME.2008.4607624

Parikh, D., Zitnick, C., Chen, T.: Determining patch saliency using low-level context. Berlin, Germany 2, 446–459 (2008)

Park, K.T., Moon, Y.S.: Automatic extraction of salient objects using feature maps. In: Preceedings of the IEEE Conference on Acoustics, Speech and Signal Processing (ICASSP), vol. 1, pp. 617–620 (2007), doi:10.1109/ICASSP.2007.365983

Perazzi, F., Krahenbuhl, P., Pritch, Y., Hornung, A.: Saliency filters: Contrast based filtering for salient region detection. In: Preceedings of the IEEE Conference on Computer Vision and Pattern Recognition (CVPR), pp. 733–740 (2012), doi:10.1109/CVPR.2012.6247743

Peters, R., Itti, L.: Beyond bottom-up: Incorporating task-dependent influences into a computational model of spatial attention. In: Preceedings of the IEEE Conference on Computer Vision and Pattern Recognition (CVPR), pp. 1–8 (2007), doi:10.1109/CVPR.2007.383337

Peters, R.J., Itti, L.: Applying computational tools to predict gaze direction in interactive visual environments. ACM Transactions on Applied Perception 5(2):9, 1–19 (2008), doi:10.1145/1279920.1279923

Peters, R.J., Itti, L.: Congruence between model and human attention reveals unique signatures of critical visual events. In: Advances in Neural Information Processing Systems (NIPS), Vancouver, BC, Canada (2009)

Ramanathan, S., Katti, H., Sebe, N., Kankanhalli, M., Chua, T.-S.: An eye fixation database for saliency detection in images. In: Daniilidis, K., Maragos, P., Paragios, N. (eds.) ECCV 2010, Part IV. LNCS, vol. 6314, pp. 30–43. Springer, Heidelberg (2010)

Rapantzikos, K., Tsapatsoulis, N., Avrithis, Y., Kollias, S.: Bottom-up spatiotemporal visual attention model for video analysis. IET Image Processing 1(2), 237–248 (2007)

Russell, B.C., Torralba, A., Murphy, K.P., Freeman, W.T.: Labelme: A database and web-based tool for image annotation. International Journal of Computer Vision 77(1-3), 157–173 (2008), doi:10.1007/s11263-007-0090-8

Shic, F., Scassellati, B.: A behavioral analysis of computational models of visual attention. International Journal of Computer Vision 73(2), 159–177 (2007), doi:10.1007/s11263-006-9784-6

Tatler, B.W., Baddeley, R.J., Gilchrist, I.D.: Visual correlates of fixation selection: effects of scale and time. Vision Research 45(5), 643–659 (2005), doi:10.1016/j.visres.2004.09.017

Wang, W., Wang, Y., Huang, Q., Gao, W.: Measuring visual saliency by site entropy rate. In: Preceedings of the IEEE Conference on Computer Vision and Pattern Recognition (CVPR), pp. 2368–2375 (2010), doi:10.1109/CVPR.2010.5539927

Yu, Z., Wong, H.S.: A rule based technique for extraction of visual attention regions based on real-time clustering. IEEE Transactions on Multimedia 9(4), 766–784 (2007), doi:10.1109/TMM.2007.893351

Zhai, Y., Shah, M.: Visual attention detection in video sequences using spatiotem-
    poral cues. In: Proceedings of the 14th Annual ACM International Conference
    on Multimedia, MULTIMEDIA 2006, pp. 815–824. ACM, New York (2006),
    doi:10.1145/1180639.1180824
Zhao, Q., Koch, C.: Learning a saliency map using fixated locations in natural scenes.
    Journal of Vision 11(3):9, 1–15 (2011), doi:10.1167/11.3.9

Chapter 3
Location-Based Visual Saliency Computation

This Chapter reviews the bottom-up visual saliency models for computing location-based saliency. These models can be roughly categorized into three domains, including the spatial domain, the transform domain and the spatiotemporal domain. For each domain, we will present the technical details of one or two representative approaches, while their followers and other approaches in the domain will also be briefly introduced. Note that we only focus on the bottom-up models for location-based saliency computation in this Chapter. The object-based saliency models will be discussed in Chap. 4, while the learning-based saliency models that also consider the influences of top-down factors will be presented in Chaps. 5, 6 and 7.

3.1 Overview

One of the key questions before discussing the bottom-up saliency models is to distinguish them from the top-down ones. Actually, it is often not easy to clearly distinguish them since almost all models adopt the bottom-up information processing mechanism, while top-down factors, either learned or manually fine-tuned, are progressively incorporated in certain stages to modulate the information processing. In particular, top-down models often utilize the machine learning algorithms, but the usage of learning algorithm is also insufficient to classify models into bottom-up or top-down since the learning can be used either in feature selection stage (i.e., preattentive stage) or in feature integration stage (i.e., focused attention stage).

In this book, we adopt the following criteria to distinguish various bottom-up and top-down saliency models:

- **Bottom-up Model**. A bottom-up model computes visual saliency by integrating preattentive features without any bias and the feature integration is independent of cues from predefined tasks or prior knowledge.

- **Top-down Model.** In a top-down model, the "feature-saliency" mapping is mainly guided by cues from volitional tasks or mandatory priors learned from training scenes.

From the definitions above, the main difference between bottom-up and top-down models lies in **whether cues from the volitional tasks or learned priors are taken into account during the feature integration stage.** Following this standard, we can see that models that use high-level cues such as face detector (Cerf et al, 2009), object detector (Cheng et al, 2011) or even learned sparse basis functions (Bruce and Tsotsos, 2005) are all classified into the bottom-up category since these priors take effect only in the feature selection stage. In the feature integration stage, these features will be combined without any bias to pop-out salient visual subsets. Meanwhile, models such as (Navalpakkam and Itti, 2007; Kienzle et al, 2007; Judd et al, 2009; Zhao and Koch, 2011) are all top-down models since the prior knowledge is learned from training scenes and such priors are used to guide the biased integration of various preattentive features.

In the rest part of this Chapter, we will divide existing bottom-up saliency models into several categories and describe the technical details for one or two representative models in each category. Note that only location-based bottom-up saliency models are discussed in this Chapter, in which visual saliency values are assigned to various spatial locations such as pixels, voxels or macro-blocks. With these models in mind, readers can easily capture the main branches of state-of-the-art researches in bottom-up saliency computation. The source codes for all these models are publicly available on the Internet, which can be easily downloaded and configured for testing. In this Chapter, the benchmark proposed in (Bruce and Tsotsos, 2005), which contains 120 color images and human fixations, will be used to demonstrate the performance of bottom-up saliency models that are based on images. For the video saliency models, the video fixation benchmark proposed in (Itti, 2004, 2008) will be used for performance evaluation. These two benchmarks, for the sake of convenience, are denoted as **Toronto-120** and **ORIG** in the rest of the Chapter, respectively.

3.2   Saliency Models in Spatial Domain

When computing image saliency, a fundamental hypothesis is that a salient visual subset is unique or rare and such rarity makes it pop-out easily from its surroundings. Tatler et al (2005) ever conducted eye-tracking experiments to compare the difference between fixated and non-fixated locations. They reported that high spatial frequencies are far more discriminatory than low spatial frequencies for contrast, edge-content and chromaticity. This finding was also consolidated by some subjective experiments conducted in (Elazary and Itti, 2008; Marat et al, 2009). Following this hypothesis, saliency models in spatial domain usually compute visual saliency by detecting the irregularity in the input visual stimuli, while such irregularity can be defined either locally, globally or both.

3.2.1 Saliency from Local Irregularity

A classical model for detecting local irregularity was proposed in (Itti et al, 1998). In this approach, visual saliency for a location is quantized as its difference from the neighboring locations in multiple features and multiple scales. The main framework of this approach can be summarized as three main modules, including: 1) extraction of preattentive features; 2) computation of multi-scale center-surround contrasts; and 3) integration of contrast maps.

In feature extraction module, seven preattentive features are extracted from the input visual stimulus, including intensity, red-green and blue-yellow opponencies and four orientations. Figure 3.1 demonstrates an image and its seven preattentive features. We can see that in some feature channels such as the red-green and blue-yellow opponencies, targets (green apple) can be easily distinguished from distractors (red apple). However, in feature channels such as intensity and orientations, it is infeasible to easily separate these apples. Thus a hidden assumption in (Itti et al, 1998) is that salient targets must be unique or rare in specific features and scales.

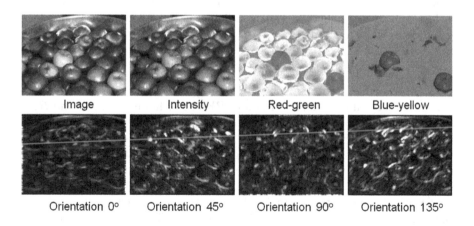

Fig. 3.1 Seven preattentive features used in (Itti et al, 1998)

In (Itti et al, 1998), the intensity feature I is calculated as:

$$I = \frac{r + g + b}{3},$$

(3.1)

where $r, g, b \in [0,1]$ are the red, green and blue channels of the input image.

Given the r, g, b channels, the red-green and blue-yellow color opponencies are extracted according to Hering's theory. In this theory, it is claimed that some colors (e.g., red and green, blue and yellow) cannot be perceived simultaneously in human vision system. Thus it is assumed that the red-green and blue-yellow color opponencies are the color information processed in our vision system. To compute the color opponencies, r, g and b are further normalized by I to decouple hue from intensity. Since subjects are not sensitive to hue variations at low luminance (e.g., $I < 0.1$), the normalized red channel $\hat{r}$ can be computed as:

$$\hat{r} = \begin{cases} r/I \text{ , if } I \geq 0.1 \\ 0 \text{ , } otherwise \end{cases}, \tag{3.2}$$

where the normalized green $\hat{g}$ and blue $\hat{b}$ can be calculated in a similar way.

Given the normalized $\hat{r}, \hat{g}$ and $\hat{b}$, four broadly-tuned colors[1] are created as:

$$R = \max(0, \hat{r} - \frac{\hat{g} + \hat{b}}{2}), \quad G = \max(0, \hat{g} - \frac{\hat{r} + \hat{b}}{2}),$$
$$B = \max(0, \hat{b} - \frac{\hat{r} + \hat{g}}{2}), \quad Y = \max(0, \hat{r} + \hat{g} - 2(|\hat{r} - \hat{g}| + \hat{b})). \tag{3.3}$$

With these color channels, the red-green and blue-yellow opponencies can be calculated as $RG = R - G$ and $BY = B - Y$, respectively. Note that in (Walther and Koch, 2006), it is argued that the definition of Y and the normalization operation in (3.3) have some drawbacks (more explains can be found in Walther, 2006). Thus a different way of computing the color opponencies was proposed:

$$RG = \frac{r - g}{\max(r, g, b)}, \quad BY = \frac{b - \min(r, g)}{\max(r, g, b)}, \tag{3.4}$$

To avoid large fluctuations of RG and BY at low luminance, RG and BY are set to zero if $\max(r, g, b) < 0.1$. Note that this definition normalize the color with $\max(r, g, b)$ instead of $(r + g + b)/3$, which is believed to be more reasonable for colors such as yellow, magenta and cyan. Actually, the two definitions only differs in the definition of "yellow" as well as the normalization operation.

The orientation features can be derived from the intensity channel by convolving it with Gabor filters in four directions. These Gabor filters can

---

[1] Note that the definitions of Y are not consistent in (Itti et al, 1998) and their source code. In (Itti et al, 1998), $Y = \max(0, (\hat{r} + \hat{g})/2 - |\hat{r} - \hat{g}|/2 - b)$, while in source code $Y = \max(0, (\hat{r} + \hat{g}) - 2(|\hat{r} - \hat{g}| + \hat{b}))$. According to some other related works, we adopt the latter definitions in this book. With this definition, R, G, B and Y fall in the same dynamic range of $[0,3]$.

approximate the receptive field impulse response of orientation-selective neurons in primary visual cortex. The convolution can be conducted as:

$$O(\theta) = |I * G_0(\theta)| + |I * G_{\pi/2}(\theta)|, \qquad (3.5)$$

where $\theta \in \{0°, 45°, 90°, 135°\}$ is the direction and

$$G_\Psi(x, y, \theta) = \exp\left(-\frac{x'^2 + \gamma^2 y'^2}{2\delta^2}\right) \cos\left(2\pi\frac{x'}{\lambda} + \Psi\right) \qquad (3.6)$$

is a Gabor filter with phase Ψ, aspect ratio γ, standard deviation δ and wavelength λ. Coordinates (x', y') are transformed from (x, y) with respect to θ:

$$\begin{aligned} x' &= y \sin(\theta) + x \cos(\theta), \\ y' &= y \cos(\theta) - x \sin(\theta). \end{aligned} \qquad (3.7)$$

Given these preattentive features, a feasible solution to detect salient targets could be popping-out the candidate targets in each feature and then combining (or selecting) the effective features. In this process, local contrast can be used as an effective cue to quantize the difference between a region and its surroundings. Recall that the bipolar and ganglion cells in retina can form two logic structures, including the on-center and the off-center. These two structures can be used to extract the most informative visual cues from the image. As shown in Fig. 3.2, such "center-surround" characteristic can be simulated by the Difference of Gaussian function. Moreover, such "center" and "surround" Gaussians should take different σ to extract the local irregularities from different scales.

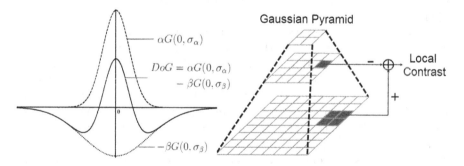

Fig. 3.2 The on-center neural structure can be approximated by Difference of Gaussian and pixel-wise subtraction across Gaussian pyramids

However, the direct computation of the average responses in various center and surround regions is very time-consuming. As an efficient approximation,

Itti et al (1998) proposed to establish a dyadic Gaussian pyramid for each preattentive feature by progressively applying low-pass filter and sub-sampling the input image by a factor of 2 (as shown in Fig. 3.2). This pyramid can be used to approximate the multi-scale Gaussian convoluted images. When using the pyramid, the average "center" response is taken from the corresponding pixel at fine scale $c \in \{2, 3, 4\}$, while the average "surround" response corresponds to the pixel at coarse scale $s = c + \delta$ with $\delta \in \{3, 4\}$. In this way, the multi-scale center-surround contrasts can be efficiently approximated by the pixel-wise subtraction.

For the 7 preattentive features and 6 center-surround scales, 42 contrast maps are computed in total: 6 for intensity, 12 for color, and 24 for orientation. Figure 3.3 demonstrates the multi-scale center-surround contrast maps extracted from an image. We can see that the local contrasts can well reveal the locations of salient targets from multiple scales. Note that the center-surround contrast maps from different features and scales may have different resolutions and dynamic ranges. To address this problem, all the maps are resized to the resolution at the 4th scale and a normalization operation $\mathcal{N}(\cdot)$ is proposed to suppress maps with numerous comparable peak responses. In the normalization, a map is first normalized to a fixed range $[0, M]$. After that, the average response at all local maxima weaker than M are computed as $\overline{m}$. Finally, the map is multiplied by $(M - \overline{m})^2$. In the normalization, maps with a distinct candidate target will be enhanced more than those with multiple candidates (i.e., a map with obvious foreground and clean background is preferred).

With $\mathcal{N}(\cdot)$, contrast maps from each of the seven preattentive features are first normalized and combined across scales. The combination results are further normalized and summed up to generate three conspicuity maps for intensity, color and orientation. Finally, these three conspicuity maps are normalized and combined linearly with equal weights to generate the final saliency map.

In Fig. 3.4 we demonstrate several representative saliency maps generated by (Itti et al, 1998) when using the **Toronto-120** benchmark. The best and worst cases are selected from the top ten and bottom ten images according to the **AUC** scores, respectively. From Fig. 3.4, we can see that the approach in (Itti et al, 1998) performs the best on small or medium targets and clean scenes. However, when salient targets are large or scenes are complex, this approach may fail since many distractors also have strong local contrasts. These contrasts may appear across scales, which can inhibit the real targets after the across-scale combination. Thus it is necessary to further compare the highly-contrasted targets and distractors which may be not adjacent to each other. Toward this end, the global irregularity detection approaches have been proposed, which will be discussed later.

To sum up, the local center-surround contrast can well simulate the neurobiological characteristics of human receptive visual field. Therefore, many saliency models in the spatial domain have adopted similar architecture.

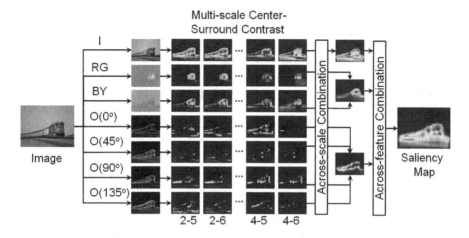

Fig. 3.3 System framework of the approach in (Itti et al, 1998)

Fig. 3.4 Representative saliency maps obtained by (Itti et al, 1998)

For example, Hu et al (2005a) used the texture difference between center region and surrounding region as the saliency criteria. Gao et al (2009) proposed a discriminant center-surround hypothesis to compute visual saliency using the local mutual information. This hypothesis equates the saliency of each image location to the discriminant power of a set of features with respect to the classification problem that opposes stimuli at center and surround, at that location.

Beyond the center-surround property, Le Meur et al (2006) proposed a bottom-up saliency model that simulated several characteristics of human vision system. They first conducted several eye-tracking experiments to infer the probable mechanisms in human vision system such as center-bias and coverage. These findings, together with some other characteristics of human vision system such as contrast sensitivity functions, perceptual decomposition, visual masking, and center-surround interactions, are then simulated to build the saliency model. Riche et al (2012) proposed to extract multi-scale rarities from the YCbCr color space using multiple Gabor filters. Kadir and Brady (2001) discussed the inherent relationship between saliency, scale and content description. They also proposed a multi-scale saliency model based on the local entropy. Bogdanova et al (2008) further extended the traditional center-surround framework to omnidirectional images. Compared with the classic image saliency model, their model is based on sphere geometry (e.g., the spherical Laplacian Pyramid and filtering algorithm) and can well adapt to the omnidirectional images with severe distortion.

One main problem of the local irregularity is that salient targets must be locally unique or rare, but the unique or rare visual subsets may not be always salient. Therefore, recent studies often take the multi-scale contrasts as candidate cues for saliency subsets, while some other cues such as global irregularity and top-down factors are further incorporated to improve the performance of visual saliency model.

3.2.2   Saliency from Global Irregularity

The major difference between local and global irregularities lies in the definitions of spatial context, from which visual cues can be extracted to determine whether a candidate location is salient or not. For local irregularity, the spatial context is defined as a surrounding region around the specific location, while the global irregularity extends such spatial context to the whole scene. For example, the simplest way to detect the global irregularity is to use the color statistics in input image and assign high saliency values to rare pixels (Cheng et al, 2011). Such simple approach works well when only one dominant salient object appears in the image but may fail in complex scenes. Thus one problem arises: how to efficiently conduct the global competition between various targets and distractors in a complex scene?

To address this problem, Harel et al (2007) proposed a graph-based approach for visual saliency modeling. The main idea of this approach comes from intuitive observations. As shown in Fig. 3.5, an image can be represented as a fully-connected graph whose nodes correspond to various visual subsets (e.g., macro-blocks) in the image and edges are weighted according to the dissimilarities and closeness between subsets. On the graph, a random walker is used whose transition from one node to another is controlled by the dissimilarity. During the random walking process, less visited nodes can progressively pop-out since they are unique or rare in the global context.

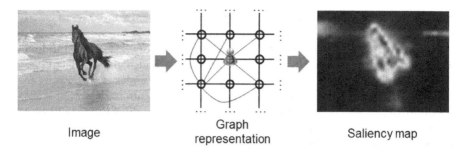

Image   Graph representation   Saliency map

Fig. 3.5 Image is represented as a graph to detect less-visited salient nodes

When calculating visual saliency using the graphical model, three major steps are required, including: 1) feature extraction; 2) graph construction; and 3) saliency computation. Among these three steps, the first step can be easily accomplished by extracting the average preattentive features in each macro-block. These features can be intensity, color, orientation, or even temporal features such as motion and flicker. As a fast approximation, the input image can be down-sampled and each macro-block corresponds to a pixel in the down-sampled image and thus can be represented by the preattentive features at that pixel. Here we assume that there are totally K preattentive features $\{\mathcal{F}_k\}_{k=1}^{K}$. When considering only the kth feature, the dissimilarity between visual subsets located at (i, j) and (p, q) can be defined as:

$$d_k((i,j)||(p,q)) \triangleq \left| \log \frac{\mathcal{F}_k(i,j)}{\mathcal{F}_k(p,q)} \right|. \tag{3.8}$$

Note that the definition of dissimilarity is actually undirected. In some cases, the logarithmic dissimilarity can be replaced with $|\mathcal{F}_k(i,j) - \mathcal{F}_k(p,q)|$, which is simpler but with comparable performance. Moreover, the closeness of two patches should also be considered in computing the edge weight $\omega((i,j),(p,q))$ by incorporating an exponential weighting schema:

$$\omega_k((i,j),(p,q)) \triangleq d_k((i,j)||(p,q)) \cdot \exp\left(-\frac{(i-p)^2 + (j-q)^2}{2\sigma^2}\right), \tag{3.9}$$

where σ is a free parameter which can be set to approximately 1/10 to 1/5 of the map width. Finally, an image can be represented by a (undirected) fully-connected graph whose nodes correspond to visual subsets and edges are weighted by dissimilarities and closeness. On this graph, a Markov chain can be defined by normalizing the weights of all outbound edges of each node to have a sum of 1. In this manner, nodes can be viewed as states and edge weights are state transition probabilities. Suppose a random walker is walking on the graph, the fraction of time he would spend at each state can well represent the node saliency. It is natural since an irregular node (visual subset) will be dissimilar from all the other nodes, which will be less visited in the random walking.

Note that it will be extremely time-consuming to simulate a real random walker on the graph. Actually, the time cost on each node can be reflected by the equilibrium distribution of the Markov chain. Since nodes in the graph are strongly connected and it is possible to pass from any state to any other state in finite steps, the Markov chain is ergodic and its equilibrium distribution exists. As a matter of fact, the equilibrium distribution can be computed by iteratively multiplying the Markov matrix with initially uniform vector and the principal eigenvector of the matrix can be generated after a small number of iterations.

After calculating the equilibrium distribution, each node is assigned a real value, which indicates the global irregularity of the corresponding visual subset. With these values, an activation map can be generated to reveal the salient locations (as shown in Fig. 3.6). However, this map may wrongly pop-out some distractors which are also unique or rare. Note that the afore-mentioned approach can concentrate saliency on limited targets and/or distractors from all candidate locations in an image, such model can be used to further suppress the noise in the activation map. That is, a new Markov chain can be built by using the value at the activation map as a new feature, while the new equilibrium distribution can be then inferred to update the activation map. As shown in Fig. 3.6, by iteratively building new graphical model and updating the activation map, energy will progressively converge to limited foreground targets, making the background much cleaner. However, we can also find that more iterations can generate cleaner background, but the performance may even become worse after several iterations (e.g., the 5th and the 6th iterations). Therefore, such operations are often conducted 2-4 times to ensure the overall performance.

Note that all the aforementioned operations are conducted on a single scale from each of the K preattentive features and multiple activation maps will be obtained from various features and scales. Therefore, the final saliency map can be generated by directly summing up all these activation maps. Some representative saliency maps on the **Toronto-120** benchmark can be found in Fig. 3.7. The best and worst cases are selected from the top ten and bottom ten images according to the **AUC** scores, respectively. From Fig. 3.7, we can see that (Harel et al, 2007) performs well when there exist compact salient regions in the scene. Since the dissimilarity scores between such regions and other distractors are very high, these regions can become less visited as a whole. However, for scenes with large targets and trivial distractors, the approach may fail since the large foreground targets may be visited more than the trivial distractors in the random walking, especially when the scene is complex (e.g., in the natural outdoor scene).

Inspired by the approach in (Harel et al, 2007), many studies have extended the graphical framework by incorporating new cues. For example, Gopalakr-ishnan et al (2009) proposed to combine the local and global

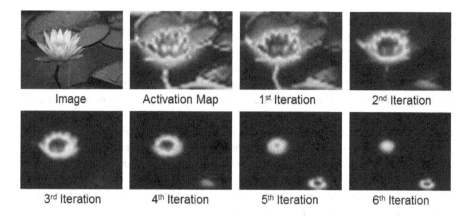

Fig. 3.6 Saliency can be concentrated to targets by iteratively building graphical models and updating activation maps

Fig. 3.7 Representative saliency maps obtained by (Harel et al, 2007)

irregularities of the image by using two graphs. A k-regular graph was used to model the local properties, while a complete graph was adopted for representing the global properties. These two graphs were then combined to identify the salient parts of the image. Cerf et al (2009) exploited the feasibility of incorporating the influence of human face into the graph-based saliency model in (Harel et al, 2007). Wang et al (2010) adopted a sparse coding framework, in which several basis functions were learned from randomly collected image patches in an unsupervised manner. Subsequently, images were projected onto the subspaces defined by these basis functions and a fully-connected graph was constructed using the projection coefficients. On this graph, the random walking was adopted to simulate the information transmission between neurons. Consequently, visual saliency was measured as the average information transmitted from one node to all the other nodes.

Beyond the graph-based approaches, some researchers tried to combine both irregularities to simultaneously estimate visual saliency from the local and global perspectives. Goferman et al (2010) proposed a context-aware saliency model which incorporated the local contrasts, global rarity and high-level factors. Vikram et al (2012) proposed to extract a set of random rectangles from the whole image. Salient pixels in each rectangle were then detected according to local statistics in Lab color space. The local saliency values in random windows distributed in the whole scene were then combined to generate the final saliency map. Borji and Itti (2012) extracted the patch-wise local and global irregularities from RGB and LAB color spaces. In their approach, image was divided into non-overlapping patches and each patch was then represented by the linear coefficients on the sparse codes learned from natural scenes. In the subspace formed by the sparse codes, local saliency was computed as the difference between center and surround patches, while global saliency was derived according to the patch's probability of happening over the entire image. Finally, the local and global saliency values from both color space were combined for image saliency estimation.

To sum up, the local and global irregularities are effective in detecting the salient targets. In most cases, targets can well pop-out from the image, if the right feature and the right scale are used. However, it is infeasible to know the best feature and scale before computing visual saliency. When using a large pool of candidate features and scales, improper features may wrongly pop-out certain distractors, while improper scale may wrongly suppress part of the large targets. Therefore, feature/scale selection is believed to be one of the most important issues in computing bottom-up saliency, which is also the major concern in learning-based top-down saliency modulation.

3.3   Saliency Models in Transform Domain

Usually, salient visual subsets are believed to be unique or rare, and such rarities usually correspond to the high spatial frequencies of the input visual signals. Therefore, a feasible solution for visual saliency modeling is to detect the irregular signals in the transform domain. Note that there are two kinds of transform domains, including the frequency domain and subspaces. In the frequency domain, visual irregularities can be effectively detected from a global perspective. In the subspaces, the input RGB signals are first projected into the predefined or learned subspaces which are believed to have better ability to pop-out the irregularities from the background.

3.3.1   Saliency from Frequency Analysis

One of the most famous works for computing saliency from frequency analysis was proposed in (Hou and Zhang, 2007). This work is motivated by the hypothesis that natural images contain rich redundancies. From the perspective of efficient coding, the redundancies are removed at the earlier stages of information processing and only unexpected signals are delivered to the later stages. Such unexpected signals, which are correlated with the "novelty" in the input scene, are considered to be salient. Thus the key problem is, given an image, how to separate the unexpected signals (i.e., novelty) from redundancies (i.e., invariants).

To address this problem, one has to figure out first what is invariant in natural image statistics. According to the studies in (Srivastava et al, 2003), there exists a power law for image statistics. That is, the power $P(\omega)$ decays as $\frac{A}{|\omega|^{2-\eta}}$, where $|\omega|$ is the magnitude of the spatial frequency ω and η changes with the image types but is usually a small number. Therefore, it is safe to assume that the energy distribution at various frequencies stay almost invariant on massive images. Moreover, the power law is a manifestation of the scale invariant nature of images, which indicates that image statistics stay invariant when scaling up or down.

Inspired by these two findings, Hou and Zhang (2007) proposed to compute visual saliency in three major steps: 1) transforming spatial signals into frequency domain to get the log amplitude spectrum; 2) removing the invariance in the log amplitude spectrum to obtain the spectral residual; and 3) transforming the spectral residual back to the spatial domain to locate the salient visual subsets. Details of the three steps are explained as follows.

Let I be the intensity channel of the input image and $\mathcal{F}[\cdot]$ be the Fourier transform, the image can be first down-sampled to 64×64 according to the scale invariant nature of images. After that, two spectra can be obtain from the down-sampled image:

$$\mathcal{A}(\omega) = |\mathcal{F}[I]|, \quad \mathcal{P}(\omega) = \varphi\left(\mathcal{F}[I]\right), \tag{3.10}$$

where $\mathcal{A}$ and $\mathcal{P}$ are the amplitude and phase spectra of the image, respectively. In this study, $\mathcal{A}(\omega)$ is further turned into the log spectrum $\mathcal{L}(\omega) = \log\left(\mathcal{A}(\omega)\right)$.

Given the log spectrum, spectral residual can be derived by removing the invariant components from it. To obtain the invariance in frequency domain, one feasible solution is to average the log spectra of massive images to get the pow distribution on frequency. However, by observing that the log spectra of various images often share similar shapes, Hou and Zhang (2007) proposed to approximate the averaged log spectrum by convolving $\mathcal{L}(\omega)$ with a 3×3 averaging filter $h(\omega)$ (as shown in Fig. 3.8). Consequently, the spectral residual can be computed as the difference between the log spectrum and its smoothed version:

$$\mathcal{R}(\omega) = \mathcal{L}(\omega) - h(\omega) * \mathcal{L}(\omega). \tag{3.11}$$

Note that the phase spectrum of the image stays unchanged. After removing the redundancies in the amplitude spectrum, the saliency map S can be generated through the inverse Fourier transform $\mathcal{F}^{-1}[\cdot]$:

$$S = G(0, \sigma) * \mathcal{F}^{-1} \left[\exp\left(\mathcal{R}(\omega) + \mathcal{P}(\omega)\right)\right]^2 , \tag{3.12}$$

where $G(0, \sigma)$ is a Gaussian filter with $\sigma = 8$ to smooth the saliency map to get better viewing experience. Some representative saliency maps on the **Toronto-120** benchmark can be found in Fig. 3.9. The best and worst cases are selected from the top ten and bottom ten images according to the **AUC** scores, respectively. From Fig. 3.9, we can see that (Hou and Zhang, 2007) also performs well on small targets, especially for those that are distinct from their surroundings. However, it is often difficult to directly modulate the saliency values of specific spatial locations in the frequency domain. Moreover, unexpected strong responses on distractors may arise near image boundaries. Furthermore, the smooth regions inside a large complex target may be wrongly suppressed since they corresponds to the low frequency in the spectrum. Therefore, post-processing operations such as border cut, non-maximum suppression and Gaussian convolution may help to further improve the overall performance. Another drawback is that this approach ignores the influence of color information, which is an important cue for visual saliency computation.

Inspired by the idea in (Hou and Zhang, 2007), Guo et al (2008) proposed to detect the spatiotemporal irregularities using the phase spectrum of quaternion Fourier transform. They observed that when reconstructing the image with only the phase spectrum, locations with less periodicity or less homogeneity will pop-out. Moreover, they propose to represent an input image as a quaternion image and the four components are intensity, red-green opponency, blue-yellow opponency and motion. Quaternion Fourier transform, and its inverse transform are then conducted on the quaternion image to derive the irregularities only from the phase spectrum. One advantage of (Guo et al, 2008) is that it can simultaneously handle multiple features in

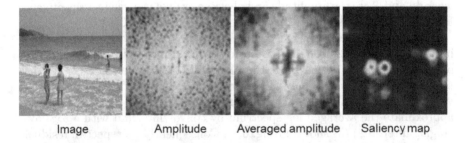

Image   Amplitude   Averaged amplitude   Saliency map

Fig. 3.8 Visual saliency can be derived from the difference between original and averaged amplitude spectra

Fig. 3.9 Representative saliency maps obtained by (Hou and Zhang, 2007)

a uniform framework. In this manner, the temporal features such as motion can be incorporated for video saliency computation.

By comparing the two approaches, we can see that Hou and Zhang (2007) emphasized the contribution of spectral residual while Guo et al (2008) argued that the main contribution was from the phase spectrum. Thus an interesting question may arise: how to distinguish the contributions from spectral residual and phase spectrum? Toward this end, we conduct a simple experiment to see the performance of (Hou and Zhang, 2007) when using only the phase spectrum in the inverse Fourier transform (i.e., set $\mathcal{R}(\omega)=0$ in (3.12)). In the evaluation, the **Toronto-120** is used as the benchmark and non-fixated pixels are re-sampled according to the flowchart shown in Fig. 2.9. We find that when using both the spectral residual and the phase spectrum, the **AUC** score can reach 0.763. On the contrary, using only the phase spectrum will make the **AUC** score decrease to 0.757. Therefore, we can conclude that both the spectral residual and phase spectrum contribute to the detection of saliency subsets, although the contribution from spectral residual is not as significant as expected.

3.3.2   Saliency from Subspace Analysis

One drawback of the frequency-based saliency analysis is the one-to-all mapping. That is, a point in frequency domain is correlated with all the visual subsets in the spatial domain. This characteristic makes it very difficult to modulate specific visual subsets while ignoring the others. One feasible solution for this problem is to perform the one-to-one mapping from the spatial domain to some subspaces. That is, a visual subset (e.g., a pixel or a macro-block) is mapped into a point in the predefined or learned subspace, in which targets and distractors can be separated easier than in the original space.

The hypothesis that different subspaces have different capabilities in distinguishing targets from distractors comes from some intuitive observations: YCbCr and Lab color spaces often outperform the RGB color space in visual saliency computation. Inspired by this hypothesis, Hu et al (2005b) proposed to model visual attention/saliency in polar 2-D space. In their approach, image was represented in a 2D space using polar transformation of its features. Thus pixels in each region fell in a 1D linear subspace. These subspaces were estimated from the data points and visual saliency was measure by projecting data points onto the normal to the subspace. The performance of this algorithm is very impressive. However, whether the polar subspace performs the best in visual saliency estimation still need to be further verified.

Beyond the predefined subspace, some approaches proposed to learn the optimal subspace in an *unsupervised* manner. For these approaches, a common process is to train a set of sparse codes (or independent components, basis functions, dictionaries, visual words, etc.) from natural image statistics (Bruce and Tsotsos, 2005; Hou and Zhang, 2009; Zhang et al, 2008; Wang et al, 2010; Borji and Itti, 2012; Sun et al, 2012). Figure 3.10 demonstrates an example of the learned sparse codes. Image patches can be thus represented as a linear combination of these sparse codes. When the sparse codes are independent components, this process can be viewed as projecting the image patch onto a subspace formed by these sparse codes. Correspondingly, a compact visual representation can be obtained for each image patch by using the projection coefficients. In this process, sparse codes act as simple-cell receptive fields similar to those appearing in the primary visual cortex of primates. Given the sparse codes, the preceding saliency computation may be realized in the context of a biological plausible neural circuit.

Actually, the main advantage of using sparse codes is to process high-dimensional signals. Suppose that a visual subset is a 8×8 macro-block, we can represent it with $D = 8 \times 8 \times 3 = 192$ RGB features. Two problems may arise due to the high dimensionality. First, the patches from an image are extremely sparse in the $\mathbb{R}^D$ space, making it difficult to perform irregularity analysis. Second, there exist strong redundancy among all the D feature dimensions, while such redundancy may misguide the saliency computation. Therefore, the main objective of using sparse codes can be described as projecting the high-dimensional signals into some subspaces for distinguish targets from distractors using compact signal representations.

Following this idea, Bruce and Tsotsos (2005) proposed an information maximization principle to measure image saliency. The system framework is shown in Fig. 3.11. In their approach, a large number of 7×7 RGB patches (360,000 in the experiments) are randomly drawn from 3,600 natural images.

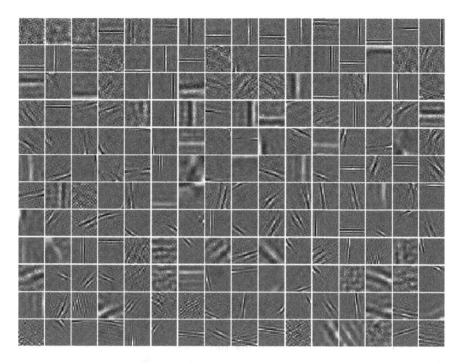

Fig. 3.10 Sparse codes trained on massive randomly selected image patches

From these images, a set of independent components are trained using the ICA algorithm. These independent components, or namely sparse codes, are used to form a subspace with much lower dimensionality (e.g., $\mathbb{R}^{147} \to \mathbb{R}^{25}$). More importantly, these sparse codes are independent from each other, making it feasible to estimate the probability distribution separately along each subspace dimension using the non-parametric density estimation. Thus the rarity of an image patch can be quantized according to the joint likelihood in the subspace, which can be derived by independently considering the likelihood of coefficients on each sparse code.

Suppose that there are totally N sparse codes, the projection from each local neighborhood of the image to these codes provides a vector w consisting of N variables w_i. For location (m, n), the variable w_i has value $v_i(m, n)$. Note that using independent codes learned from image statistics will make the components w_i (almost) independent. After the projection, visual saliency can be defined based on the strategy for maximum information sampling. That is, visual saliency for location (m, n) is computed according to the self-information:

$$s(m, n) \propto -\log(p(w_1 = v_1(m, n), \ldots, w_N = v_N(m, n))). \qquad (3.13)$$

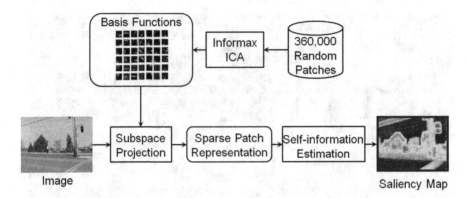

Fig. 3.11 System framework of the approach in (Bruce and Tsotsos, 2005)

According to the independence assumption on sparse codes, we have:

$$p(w_1 = v_1(m, n), \dots, w_N = v_N(m, n)) = \prod_{i=1}^{N} p(w_i = v_i(m, n)). \quad (3.14)$$

From (3.14), we can see that using the sparse representation allows the estimation in the N-D space described by w to be derived from N 1-D probability density functions. Thus the key problem turns to estimate the density function from the projection coefficients on each sparse code.

Without loss of generality, Bruce and Tsotsos (2005) proposed to estimate the density using Gaussian kernels. Let $\{v_i(s, t), (s, t) \in \Psi\}$ be the set of coefficients in the neighboring context Ψ centered at (m, n), an estimate of $p(w_i = v_i(m, n))$ in the context Ψ is given by using a Gaussian window:

$$p(w_i = v_i(m, n)) = \frac{1}{\sqrt{2\pi}\sigma} \sum_{(s,t) \in \Psi} \omega_{st} \exp\left(-\frac{(v_i(m, n) - v_i(s, t))^2}{2\sigma^2}\right), \quad (3.15)$$

where ω_{st} describes the contribution of the coefficient located at $(s, t) \in \Psi$ to the probability estimation and $\sum_{(s,t) \in \Psi} \omega_{st} = 1$.

From (3.14) and (3.15), we can see that the definition of visual saliency in (Bruce and Tsotsos, 2005) is actually derived from the center-surround difference. That is, visual saliency is measured independently from each dimension of the subspace by calculating the difference between the projection coefficients of each center region and its surrounding regions. Note that such difference is projected into [0,1] using the exponential operator, which is further re-weighted with a Gaussian function. Finally, the center-surround difference values from various locations and feature dimensions are combined to compute the final saliency. This also explains why the saliency maps computed by this approach on the benchmark **Toronto-120** are often very noisy,

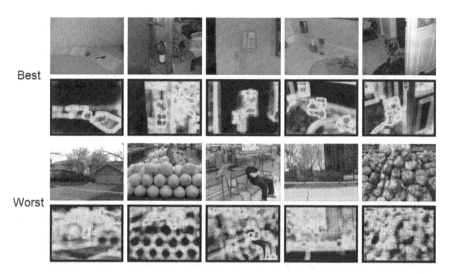

Fig. 3.12 Representative saliency maps obtained by (Bruce and Tsotsos, 2005)

even for those with the best performance. As shown in Fig. 3.12, most locations will be assigned a positive saliency value with the additive framework of the exponential terms. In this manner, the distractors in the background can also pop-out along with the salient targets.

Similarly, some other subspace-based approaches have been proposed in recent years. For example, Hou and Zhang (2009) proposed the Incremental Coding Length, which was used as the criterion to redistribute the limited energy (saliency) amongst features. Borji and Itti (2012) proposed an approach to estimate visual saliency by using the projection coefficients to quantify the local center-surround difference and the global rarity. In particular, the RGB and Lab color spaces were used simultaneously for learning the sparse codes as well as computing visual saliency. In (Wang et al, 2010), a set of sub-band feature maps were first extracted using the learned sparse codes. These feature maps were then represented as fully-connected graphs, on which random walkers were used and visual saliency was defined by the average information transmitted during the random walk. Different from these approaches, Yang and Yang (2012) proposed a novel algorithm for learning the visual words. In their approach, the visual words were treated as the latent variables of Conditional Random Field (CRF). By jointly learning the CRF and the dictionary, the overall performance of the learned saliency model was greatly improved and the estimated saliency maps become much clear.

To sum up, the latent assumption in these statistical approaches is that foreground targets and background distractors are more distinguishable in the new subspace formed by the predefined subspace or learned visual words. These visual words, which are often learned unsupervisedly from thousands of images, can be viewed as some kinds of prior knowledge. However, one problem of these approaches is that visual subsets are also equally treated in the new subspace and no bias is applied. By projecting the image patches onto the new subspace, the problem of wrongly suppressing targets and improperly popping-out distractors can be mitigated but remains unsolved.

3.4   Saliency Models in Spatiotemporal Domain

Beyond the spatial saliency, some approaches focus on computing visual saliency in the spatiotemporal domain. From the perspective of signal processing, spatial saliency only focuses on the input visual signals, while spatiotemporal saliency takes all the visual signals received in a past short period into account. Compared with the spatial saliency models, the temporal factors can help to remove some distractors that are also salient in the spatial domain (e.g., tv logos and banners). Thus the spatiotemporal saliency computation is much more challenging than the pure spatial approaches.

3.4.1   Saliency from Surprise

As stated above, natural scenes contain rich redundancies, while such redundancies are removed in the early stages of visual pathway and only unexpected signals are delivered to the late stages for further processing. Furthermore, video frames contain much more redundancies than natural images due to their strong temporal correlations. Therefore, video saliency can be computed by either removing the redundancies or detecting the unexpected signals.

In (Itti and Baldi, 2005a), a Bayesian approach is proposed to detect unexpected visual signals (i.e., the surprise). In this approach, surprise is believed to exist only in the presence of uncertainty arising from intrinsic stochasticity and missing information. A deterministic and predictable real-time stimulus contains no surprises for a given observer. Therefore, a Bayesian framework is proposed to model such uncertainty. In this framework, the uncertainty is quantized according to the prior and posterior distributions of visual signals. Given a discrete information stream of 1-D visual stimulus, its probability distribution can be modeled by models M in a model space $\mathcal{M}$. Suppose that the prior probability distribution over the model M has been initialized as $P(\mathcal{M})$ by the signals ever received, thus the new observation $\mathcal{D}$ can update $P(\mathcal{M})$ by the posterior probability distribution:

$$P(M|D) = P(M)\frac{P(D|M)}{P(D)}. \tag{3.16}$$

In this framework, the new observation D carries no surprise if it leaves M unchanged (i.e., the posterior is identical to the prior). On the contrary, D is surprising if the posterior changes greatly from the prior distribution. Therefore, the "surprise" carried in observation D can be measured by the difference between prior and posterior distributions (as shown in Fig. 3.13). In (Itti and Baldi, 2005a), surprise is quantized by the Kullback-Leibler (KL) divergence between $P(M)$ and $P(M|D)$:

$$S(D, M) = KL(P(M|D), P(M)) = \int_{\mathcal{M}} P(M|D) \log \frac{P(M|D)}{P(M)} dM. \quad (3.17)$$

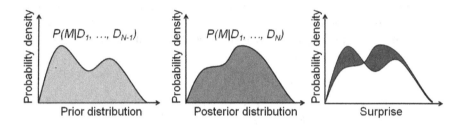

Fig. 3.13 Surprise is the difference between prior and posterior distributions

Note that KL is not symmetric and it is invariant with respect to reparameterizations. Armed with this theoretical framework, Itti and Baldi (2005b) set $\mathcal{M}$ to be a set of Poisson processes parameterized by a single λ which well describe cortical pyramidal cell firing statistics (Softky and Koch, 1993):

$$P(M(\lambda)) = \gamma(\lambda; \alpha, \beta) = \frac{\beta^{\alpha} \lambda^{\alpha-1} e^{-\beta\lambda}}{\Gamma(\alpha)}, \quad (3.18)$$

where $\alpha > 0$ and $\beta > 0$ are the shape and inverse scale parameters, respectively. $\Gamma(\cdot)$ is the Euler Gamma function. Note that the surprise is formulated on 1-D signals. Thus to detect surprise, image should be turned into a combination of 1-D information streams. Toward this end, 12 preattentive features, including intensity, red-green and blue-yellow color opponencies, four orientation, four motion energy in four directions and flicker are extracted from the input image. After that, the center-surround contrasts are computed at 6 scales (3 center scales and 2 surrounding scales) on these 12 features. Finally, 72 local contrast maps are obtained, which are then normalized to have the resolution at the 4th scale (i.e., 40×30 for an input image of 640×480). Consequently, every location of each of the 72 feature maps corresponds to a 1-D information stream for surprise detection.

To detect surprise at several time scales, five surprise detectors are implemented at every location of each feature maps. The first (fastest) detector is updated with data from feature map and the detector i samples from detector $i-1$. For an 640×480 images, there are $40 \times 30 \times 72 \times 5 = 432,000$ surprise detectors in total. Given an observation $D = \overline{\lambda}$ at a detector and the prior distribution $\gamma(\lambda; \alpha, \beta)$, the posterior distribution $\gamma(\lambda; \alpha', \beta')$ is also Gamma density, with:

$$\alpha' = \alpha + \overline{\lambda}, \quad \beta' = \beta + 1. \tag{3.19}$$

Moreover, a forgetting factor $\zeta > 0$ is introduced to prevent α and β from increasing unboundedly:

$$\alpha' = \zeta\alpha + \overline{\lambda}, \quad \beta' = \zeta\beta + 1. \tag{3.20}$$

Experimental results show that ζ can be set to 0.7. Actually, ζ can preserve the priors mean α/β but increase its variance α/β^2, embodying relaxation of belief in the prior's precision.

By incorporating (3.18) and (3.20) into the definition in (3.17), temporal surprise can be measured as:

$$
\begin{aligned}
S_T(D, M) &= KL(\gamma(\lambda; \alpha, \beta), \gamma(\lambda; \alpha', \beta')) \\
&= \alpha' \log \frac{\beta}{\beta'} + \log \frac{\Gamma(\alpha')}{\Gamma(\alpha)} + \frac{\alpha\beta'}{\beta} + (\alpha - \alpha')\Psi(\alpha),
\end{aligned}
\tag{3.21}
$$

where $\Psi(\cdot)$ is digamma function. With this model, temporal surprise can be progressively detected from the latest observation on local contrasts and the corresponding statistical priors. The temporal surprise can well reveal the most informative locations from each feature dimension. Consequently, the overall surprise can be derived by summing across features, such that a location may be surprising by its color, motion, or other features. Moreover, spatial factors can be also taken into account to further improve the overall performance, e.g., by detecting the spatial surprise (Itti and Baldi, 2005b).

Some representative saliency maps are shown in Fig. 3.14. Note that one previous frame is also demonstrated along with each frame to show the inter-frame variations. From the results, we can see that the surprise model performs the best in scenes with statistic background. In these scenes, the distractors which are also salient in the spatial domain will be suppressed in the temporal domain since they cannot provide new information. However, the location-based surprise detector has two main drawbacks. First, the computational complexity is very high. Although the updating of each model is simple, it becomes really time-consuming when the number of detectors grows up to 432,000. Second, each detector is attached to a specific location, while the correlation between detectors are beyond consideration. When encountering dynamic background (e.g., the 3rd and the 4th rows in Fig. 3.14), each detector in background region may output strong response. On the contrary, the salient target, which may be actively tracked by camera motion and stay around scene center, will be wrongly suppressed since they contain less "novelty" in the temporal axis.

Fig. 3.14 Representative saliency maps obtained by (Itti and Baldi, 2005b)

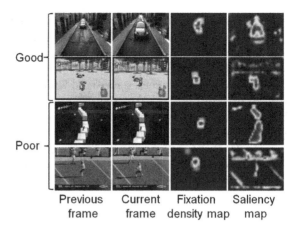

3.4.2 Saliency from Spatiotemporal Irregularity

To handle the video with dynamic background, a feasible solution is to detect irregular motion instead of using location-based surprise detector. Toward this end, Zhai and Shah (2006) proposed to detect irregular motion through point correspondence between adjacent video frames. They first located the SIFT (Lowe, 2004) points from each video frame. The RANSAC algorithm was then conducted on the inter-frame point correspondences to estimate multiple homographies that model different motion segments in the scene. Finally, the spatiotemporal saliency can be obtained by combining the temporal saliency derived from inter-frame homographies and the spatial saliency estimated by color histograms. Li and Lee (2007) estimated the motion irregularity from optical flow. Regions with irregular motion will be assigned with high temporal saliency. They also detected spatial saliency from preattentive features such as color and direction, while the center-bias effect was also modeled to compute the spatial saliency. The spatial and temporal saliency maps were then combined to generate the spatiotemporal saliency map. Marat et al (2009) presented a biology-inspired spatiotemporal saliency model. They extracted two signals from video stream corresponding to the two main outputs of the retina (i.e., parvocellular and magnocellular). Both signals were then transformed and fused into a spatiotemporal saliency map.

Another way to compute video saliency is to extend the classic saliency principle from the spatial domain to the spatiotemporal domain. For example, Itti et al extended the classic center-surround contrast framework in (Itti et al, 1998) by directly incorporating motion and flicker as the candidate preattentive features. Rapantzikos et al (2007) also extended the same center-surround framework to video saliency detection. Instead of using temporal features such as motion and flicker, they divided video into shots and built 3D feature volumes for intensity, color opponencies, 2D and 3D orientations. The center-surround difference was then computed in these 3D feature volumes, which were normalized and fused as in (Itti et al, 1998) to compute video saliency. Seo and Milanfar (2009) proposed an unified approach for both static and space-time saliency detection. This approach was also motivated by the center-surround similarity. First, a matrix consisting of local descriptors was extracted from each pixel (image) or voxel (video). The columns of the matrix represented the output of local steering kernels. Then the matrix cosine similarity was adopted to measure the resemblance of a pixel/voxel to its surroundings. Finally, the pair-wise similarities were combined to detect salient image regions or video volumes that demonstrate low similarity with their surroundings.

3.5   Notes

Generally speaking, all these location-based bottom-up approaches discussed above are motivated by one idea: salient visual subsets are unique or rare. Actually, the main difference of these approaches is from their definitions on rareness or irregularity. The pair-wise difference, rarity from spatial statistics, residual in frequency and novelty from temporal statistics are both effective to detect the salient subsets in specific scenes. However, these approaches have two main drawbacks:

The first drawback is from the mechanism of unbiased signal processing. That is, the location-based approaches often divide an input scene into multiple subsets with the same size, which are processed independently to detect the salient subsets. However, various visual subsets in a scene are often inherently correlated. For example, for a human face, the eyes and nose are tightly correlated with ears and mouth. These inherently correlated subsets should be processed as a whole in visual saliency computation. This also explains why these location-based approaches can well pop-out the small or medium targets but often fail on large targets. To solve this problem, one feasible solution is to detect salient objects, instead of salient locations. As stated in Chap. 1, locations and objects are all probable attentive units. Thus we will propose how to compute object-based saliency in Chap. 4.

The second drawback is from the fundamental motivation of these approaches. Although salient visual subsets are unique or rare, it is obvious that unique or rare subsets are not always salient. From the saliency maps obtained by these approaches, we can see that background distractors may also become unique or rare, either locally or globally. Actually, it is very difficult to suppress these distractors in unbiased bottom-up competition. In the worst cases, the real distractors are improperly popped-out while the real targets become wrongly suppressed. In human vision system, it is believed that our prior knowledge plays an important role in suppressing the unwanted distractors. Thus a computer vision system should learn such priors from training data, while the learned knowledge can be transferred to new scenes for recovering the targets that are wrongly suppressed and inhibiting the distractors that are improperly popped-out. This is also the main objective of the top-down saliency approaches that we will discuss in Chaps. 5-7.

References

Bogdanova, I., Bur, A., Hugli, H.: Visual attention on the sphere. IEEE Transactions on Image Processing 17(11), 2000–2014 (2008), doi:10.1109/TIP.2008.2003415

Borji, A., Itti, L.: Exploiting local and global patch rarities for saliency detection. In: Preceedings of the IEEE Conference on Computer Vision and Pattern Recognition (CVPR), pp. 478–485 (2012), doi:10.1109/CVPR.2012.6247711

Bruce, N.D., Tsotsos, J.K.: Saliency based on information maximization. In: Advances in Neural Information Processing Systems (NIPS), Vancouver, BC, Canada, pp. 155–162 (2005)

Cerf, M., Harel, J., Einhauser, W., Koch, C.: Predicting human gaze using low-level saliency combined with face detection. In: Advances in Neural Information Processing Systems (NIPS), Vancouver, BC, Canada (2009)

Cheng, M.M., Zhang, G.X., Mitra, N., Huang, X., Hu, S.M.: Global contrast based salient region detection. In: Preceedings of the IEEE Conference on Computer Vision and Pattern Recognition (CVPR), pp. 409–416 (2011), doi:10.1109/CVPR.2011.5995344

Elazary, L., Itti, L.: Interesting objects are visually salient. Journal of Vision 8(3):3, 1–15 (2008), doi:10.1167/8.3.3

Gao, D., Mahadevan, V., Vasconcelos, N.: The discriminant center-surround hypothesis for bottom-up saliency. In: Advances in Neural Information Processing Systems, NIPS (2009)

Goferman, S., Zelnik-Manor, L., Tal, A.: Context-aware saliency detection. In: Preceedings of the IEEE Conference on Computer Vision and Pattern Recognition (CVPR), pp. 2376–2383 (2010), doi:10.1109/CVPR.2010.5539929

Gopalakrishnan, V., Hu, Y., Rajan, D.: Random walks on graphs to model saliency in images. In: Proceedings of the IEEE Conference on Computer Vision and Pattern Recognition (CVPR), pp. 1698–1705 (2009), doi:10.1109/CVPR.2009.5206767

Guo, C., Ma, Q., Zhang, L.: Spatio-temporal saliency detection using phase spectrum of quaternion fourier transform. In: Preceedings of the IEEE Conference on Computer Vision and Pattern Recognition (CVPR), pp. 1–8 (2008), doi:10.1109/CVPR.2008.4587715

Harel, J., Koch, C., Perona, P.: Graph-based visual saliency. In: Advances in Neural Information Processing Systems (NIPS), pp. 545–552 (2007)

Hou, X., Zhang, L.: Saliency detection: A spectral residual approach. In: Proceedings of the IEEE Conference on Computer Vision and Pattern Recognition (CVPR), pp. 1–8 (2007), doi:10.1109/CVPR.2007.383267

Hou, X., Zhang, L.: Dynamic visual attention: Searching for coding length increments. In: Advances in Neural Information Processing Systems (NIPS), pp. 681–688 (2009)

Hu, Y., Rajan, D., Chia, L.T.: Adaptive local context suppression of multiple cues for salient visual attention detection. In: Preceedings of the IEEE International Conference on Multimedia and Expo, ICME (2005a), doi:10.1109/ICME.2005.1521431

Hu, Y., Rajan, D., Chia, L.T.: Robust subspace analysis for detecting visual attention regions in images. In: Proceedings of the 13th Annual ACM International Conference on Multimedia, MULTIMEDIA 2005, pp. 716–724. ACM, New York (2005), doi:10.1145/1101149.1101306

Itti, L.: Automatic foveation for video compression using a neurobiological model of visual attention. IEEE Transactions on Image Processing 13(10), 1304–1318 (2004), doi:10.1109/TIP.2004.834657

Itti, L.: Crcns data sharing: Eye movements during free-viewing of natural videos. In: Collaborative Research in Computational Neuroscience Annual Meeting, Los Angeles, California (2008)

Itti, L., Baldi, P.: Bayesian surprise attracts human attention. In: Advances in Neural Information Processing Systems (NIPS), pp. 547–554 (2005a)

Itti, L., Baldi, P.: A principled approach to detecting surprising events in video. In: Preceedings of the IEEE Conference on Computer Vision and Pattern Recognition (CVPR), vol. 1, pp. 631–637 (2005b), doi:10.1109/CVPR.2005.40

Itti, L., Koch, C., Niebur, E.: A model of saliency-based visual attention for rapid scene analysis. IEEE Transactions on Pattern Analysis and Machine Intelligence 20(11), 1254–1259 (1998), doi:10.1109/34.730558

Judd, T., Ehinger, K., Durand, F., Torralba, A.: Learning to predict where humans look. In: Proceedings of the IEEE International Conference on Computer Vision (ICCV), pp. 2106–2113 (2009), doi:10.1109/ICCV.2009.5459462

Kadir, T., Brady, M.: Saliency, scale and image description. International Journal of Computer Vision 45(2), 83–105 (2001), doi:10.1023/A:1012460413855

Kienzle, W., Wichmann, F.A., Scholkopf, B., Franz, M.O.: A nonparametric approach to bottom-up visual saliency. In: Advances in Neural Information Processing Systems (NIPS), pp. 689–696 (2007)

Le Meur, O., Le Callet, P., Barba, D., Thoreau, D.: A coherent computational approach to model bottom-up visual attention. IEEE Transactions on Pattern Analysis and Machine Intelligence 28(5), 802–817 (2006), doi:10.1109/TPAMI.2006.86

Li, S., Lee, M.C.: Efficient spatiotemporal-attention-driven shot matching. In: Proceedings of the 15th Annual ACM International Conference on Multimedia, MULTIMEDIA 2007, pp. 178–187. ACM, New York (2007), doi:10.1145/1291233.1291275

Lowe, D.G.: Distinctive image features from scale-invariant keypoints. International Journal of Computer Vision 60(2), 91–110 (2004), doi:10.1023/B:VISI.0000029664.99615.94

Marat, S., Ho Phuoc, T., Granjon, L., Guyader, N., Pellerin, D., Guérin-Dugué, A.: Modelling spatio-temporal saliency to predict gaze direction for short videos. International Journal of Computer Vision 82(3), 231–243 (2009), doi:10.1007/s11263-009-0215-3

Navalpakkam, V., Itti, L.: Search goal tunes visual features optimally. Neuron 53, 605–617 (2007)

Rapantzikos, K., Tsapatsoulis, N., Avrithis, Y., Kollias, S.: Bottom-up spatiotemporal visual attention model for video analysis. IET Image Processing 1(2), 237–248 (2007)

Riche, N., Mancas, M., Gosselin, B., Dutoit, T.: Rare: A new bottom-up saliency model. In: Preceedings of the 19th IEEE International Conference on Image Processing (ICIP), pp. 641–644 (2012), doi:10.1109/ICIP.2012.6466941

Seo, H.J., Milanfar, P.: Static and space-time visual saliency detection by self-resemblance. Journal of Vision 9(12):15, 1–27 (2009), doi:10.1167/9.12.15

Softky, W.R., Koch, C.: The highly irregular firing of cortical cells is inconsistent with temporal integration of random epsps. The Journal of Neuroscience 13(1), 334–350 (1993)

Srivastava, A., Lee, A.B., Simoncelli, E.P., Zhu, S.C.: On advances in statistical modeling of natural images. Journal of Mathematical Imaging and Vision 18, 17–33 (2003)

Sun, X., Yao, H., Ji, R.: What are we looking for: Towards statistical modeling of saccadic eye movements and visual saliency. In: Preceedings of the IEEE Conference on Computer Vision and Pattern Recognition (CVPR), pp. 1552–1559 (2012), doi:10.1109/CVPR.2012.6247846

Tatler, B.W., Baddeley, R.J., Gilchrist, I.D.: Visual correlates of fixation selection: Effects of scale and time. Vision Research 45(5), 643–659 (2005), doi:10.1016/j.visres.2004.09.017

Vikram, T.N., Tscherepanow, M., Wrede, B.: A saliency map based on sampling an image into random rectangular regions of interest. Pattern Recognition, 3114–3124 (2012)

Walther, D.: Interactions of visual attention and object recognition: Computational modeling, algorithms, and psychophysics. PhD thesis, California Institute of Technology (2006)

Walther, D., Koch, C.: Modeling attention to salient proto-objects. Neural Networks 19(9), 1395–1407 (2006)

Wang, W., Wang, Y., Huang, Q., Gao, W.: Measuring visual saliency by site entropy rate. In: Preceedings of the IEEE Conference on Computer Vision and Pattern Recognition (CVPR), pp. 2368–2375 (2010), doi:10.1109/CVPR.2010.5539927

Yang, J., Yang, M.H.: Top-down visual saliency via joint crf and dictionary learning. In: Preceedings of the IEEE Conference on Computer Vision and Pattern Recognition (CVPR), pp. 2296–2303 (2012), doi:10.1109/CVPR.2012.6247940

Zhai, Y., Shah, M.: Visual attention detection in video sequences using spatiotemporal cues. In: Proceedings of the 14th Annual ACM International Conference on Multimedia, MULTIMEDIA 2006, pp. 815–824. ACM, New York (2006), doi:10.1145/1180639.1180824

Zhang, L., Tong, M.H., Marks, T.K., Shan, H., Cottrell, G.W.: Sun: A bayesian framework for saliency using natural statistics. Journal of Vision 8(7):32, 1–20 (2008), doi:10.1167/8.7.32

Zhao, Q., Koch, C.: Learning a saliency map using fixated locations in natural scenes. Journal of Vision 11(3):9, 1–15 (2011), doi:10.1167/11.3.9

Chapter 4
Object-Based Visual Saliency Computation

This Chapter describes the approaches for object-based saliency computation, which can be roughly grouped into two categories. Approaches in the first category focus on segmenting the whole salient object by using location-based saliency maps, while approaches in the second category focus on directly computing visual saliency on object level. In this Chapter, we will introduce the technical details of six approaches from these two categories and their performance will be compared at the end of this Chapter.

4.1 Overview

In recent years, the number of digital images on the Internet has grown dramatically. In these images, the truly meaningful parts may be just a small proportion. The nontrivial contents, usually in the form of salient objects, are sufficient to represent the semantic meanings in most cases and consequently play an important role in many image applications such as content-based image retrieval and video advertising. Consequently, such requirements give rise to the studies on object-based saliency inference. Different from the location-based saliency for pixels and macro-blocks, these studies propose to detect and analyze the salient object as a whole. According to the neurobiological mechanisms discussed in Chap. 1, object is also a probable attentive unit. Therefore, the object-based saliency models are also compatible with the neurobiological mechanisms in human vision system.

Usually, there are two different categories of approaches to detect salient objects in a scene. Approaches in the first category focus on utilizing the location-based saliency map. They aim to extract salient objects from the saliency maps computed using the location-based saliency models. Approaches in the second category propose to directly measure the saliency of objects. They first divide an image into a set of objects (regions, superpixels) and then directly compute visual saliency in object level. In this Chapter, we will describe the technical details of several representative approaches from the two categories. As shown in Fig. 4.1, Sect. 4.2 will focus on detecting salient object from location-based saliency maps by assuming them to be accurate or not. In Sect. 4.3, we will propose how to directly measure object saliency by segmenting images into regions with distinct sizes or super-pixels with comparable sizes. Note that this Chapter only focuses on computing object-based saliency in images, while detecting salient objects in video is beyond the scope of this Chapter.

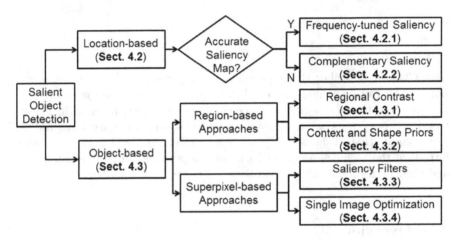

Fig. 4.1 The overall structure of the approaches introduced in Chap. 4

4.2 Salient Object Extraction from Location-Based Saliency Map

Since there already exist various saliency models that can reveal the salient locations in a scene, it is necessary to explore how to detect salient objects from location-based saliency map. In this Section, we will first discuss how to detect salient objects when assuming that the computed location-based saliency map is accurate. After that, we will propose how to detect objects from inaccurate saliency maps.

4.2.1   Frequency-Tuned Saliency

Because location-based saliency well accords with human visual perception and can be used as one sort of selection mechanisms of the important content, the location-based saliency models have been widely used in many approaches for salient object detection. Among these approaches, the most straight-forward solution is to detect the location-based saliency and then generate the salient objects through object segmentation or fuzzy growing. For example, Itti et al (1998) combined multi-scale features into a single topographical saliency map and adopted a dynamical neural network to select the attended areas that roughly contained the interesting objects. Ma and Zhang (2003) generated a contrast-based saliency map and extracted objects by fuzzy growing. Hou and Zhang (2007) first constructed the location-based saliency map by analyzing the log-spectrum of the image and then adopted a simple thresholding operation to detect salient objects.

As an representative approach in this category, Achanta et al (2009) presented a frequency-tuned approach to detect salient regions in images. They proposed that the following rules should be set for the location-based saliency detectors:

1. Emphasize the largest salient objects.
2. Uniformly highlight whole salient regions.
3. Establish well-defined boundaries of salient objects.
4. Disregard high frequencies from texture, noise and blocking artifacts.
5. Efficiently output full resolution saliency maps.

In their approach, the five saliency rules correspond to the filtering strategies for the input visual signals. Given an image, let ω_{lc} and ω_{hc} be the low and high frequency cut-off values, respectively. For the 1st and 2nd rules, ω_{lc} should be low to keep the very low frequencies from the original image. For the 3rd and 4th rules, high frequencies should be maintained to keep the clear object boundaries, while the highest frequencies should be disregarded to avoid noise, coding artifacts, and texture patterns. This indicates that ω_{hc} should not be too high. Moreover, such filtering process should be conducted in the original image resolution to output full-resolution saliency maps.

Following these filtering strategies, a set of band-pass filters with contiguous pass bands $[\omega_{lc}, \omega_{hc}]$ are adopted. In this approach, the band-pass Difference of Gaussian (DoG) filter is adopted, which is widely used in edge detection since it closely and efficiently approximates the Laplacian of Gaussian (LoG) filter for detecting intensity changes. The DoG filter is defined as:

$$DoG(x,y) = \frac{1}{2\pi}\left[\frac{1}{\sigma_1^2}e^{-\frac{x^2+y^2}{2\sigma_1^2}} - \frac{1}{\sigma_2^2}e^{-\frac{x^2+y^2}{2\sigma_2^2}}\right] = G(x,y,\sigma_1) - G(x,y,\sigma_2),$$

(4.1)

where σ_1 and σ_2 are the standard deviations of the Gaussian functions $G(x,y,\sigma_1)$ and $G(x,y,\sigma_2)$, respectively.

A DoG filter is actually a band-pass filter whose passband width is controlled by the ratio $\rho = \frac{\sigma_1}{\sigma_2}$. Consequently, a summation over DoG with standard deviations in the ratio ρ results in:

$$\sum_{n=0}^{N-1} G(x, y, \rho^{n+1}\sigma) - G(x, y, \rho^n \sigma) = G(x, y, \rho^N \sigma) - G(x, y, \sigma). \qquad (4.2)$$

That is, applying a set of continues DoG filters is equivalent to applying one DoG with standard deviations σ and $\rho^N \sigma$. Therefore, the simple difference between the coarsest and finest image representations can well maintain the frequencies in various bands. That is why the salient regions will be fully covered and not just high-lighted on edges or in the center of the regions.

In the implementation, the high frequency σ_1 is set to infinity. This results in a notch in frequency at DC while retaining all other frequencies. The low frequency is simulated by the binomial filter $\frac{1}{16}[1, 4, 6, 4, 1]$ giving $\omega_{hc} = \pi/2.75$. Correspondingly, the saliency value for a pixel (x, y) can be computed as:

$$s(x, y) = \|\mathbf{v}_\mu - \mathbf{v}_{\omega_{hc}}(x, y)\|, \qquad (4.3)$$

where $\mathbf{v}_\mu$ is the mean pixel value of the image (e.g., RGB/Lab features) and $\mathbf{v}_{\omega_{hc}}(x, y)$ is from the image smoothed by the binomial filter. With this definition, location-based saliency can be computed very efficiently by simply measuring the pixel-wise difference between an image and its smoothed version.

Given the location-based saliency, salient objects can be detected through image segmentation. In this process, images are segmented into objects with the mean-shift segmentation algorithm. The segmentation is performed in Lab color space with fixed parameters for all images. The saliency value for each object is then computed as the average saliency value of all pixels in the object. After that, an intelligent threshold is computed to find the salient objects, which is computed as:

$$T_a = \frac{2}{W \times H} \sum_{x=1}^{W} \sum_{y=1}^{H} s(x, y), \qquad (4.4)$$

where W and H are the width and height of the saliency map in pixels, respectively. Some representative examples of the detected salient objects can be found in Fig. 4.2. We can see that the performance is impressive when the location-based salience is accurate. However, for inaccurate saliency maps, some parts of the salient objects may get wrongly suppressed, leading to the integrity problem.

Fig. 4.2 Saliency maps and salient objects obtained by (Achanta et al, 2009)

4.2.2   Complementary Saliency

When the location-based saliency is precise and complete, the aforementioned approach can achieve promising results. However, existing location-based approaches often have difficulties to pop-out a complex object as a whole since they ignore the inherent correlations between these subsets. In this case, various object parts will be processed independently and the integrity problem may arise due to the improperly selected scales. Suppose that a complex object can pop-out as a whole at a specific scale, the detected salient subsets may be either a subset of the object when using a smaller scale or a large region that contains some redundant part of the background when using a larger scale. For the sake of simplification, we refer to the two cases as sketch-like and envelop-like saliency maps, respectively.

Since it is infeasible to go through all possible scales to find the best one to pop-out the complex objects, we investigate how to extract the entire salient object from the sketch-like and envelop-like saliency maps generated by location-based saliency models. The idea of extracting salient objects by using sketch-like and envelope-like saliency maps is motivated by their complementary property. That is, the sketch-like saliency map usually has high precision in locate the salient object, while the envelope-like saliency map often demonstrates high recall. By using the two complementary saliency maps for the same image, we can obtain two sets of pixels, one with reliable foreground pixels and one with confident background pixels. Then a classifier can be trained to identify the rest pixels as foreground or background. By transferring the complex object extraction task to an easier classification problem, this approach can effectively address the integrity problem. To sum up, this approach contains three main steps: 1) computing complementary saliency maps; 2) modeling foreground and background pixels; and 3) classifying pixels for salient object extraction.

Before describing the technical details of these three steps, we give two definitions first which will be used later. Suppose the pixel set of a salient object is denoted as $\mathbf{O}$, then $\mathbf{E}$ is called the *envelope* of the object if $\mathbf{O} \subseteq \mathbf{E}$, and $\mathbf{S}$ is called the *skeleton* of the object if $\mathbf{S} \subseteq \mathbf{O}$. In this study, it is allowed that $\mathbf{O} - \mathbf{E} \neq \emptyset$ and $\mathbf{S} - \mathbf{O} \neq \emptyset$ if the following two conditions are satisfied: 1) the envelope covers the interesting objects as much as possible while leaving just few redundant background areas; and 2) the skeleton contains the most representative parts of the object while including little background.

Following these definitions, two complementary saliency maps should be generated to obtain two complementary pixel sets (i.e., envelope and skeleton). Then pixels outside the two sets will be classified to detect the whole object. The framework of the proposed approach is shown in Fig. 4.3. To generate the complementary saliency maps, two location-based saliency maps are computed first. The first is frequency-tuned saliency map (**FSM**) proposed in (Achanta et al, 2009), and the second is center-surround contrast map (**CCM**) derived from (Liu et al, 2007b). The reason of choosing these two feature maps is that **FSM** can output desirable results with very efficient computation while **CCM** can well represent the regional contrast feature and is insensitive to local sudden changes. Besides, as we will illuminate later, **CCM** fails near the image boundary, but **FSM** can work well for the border.

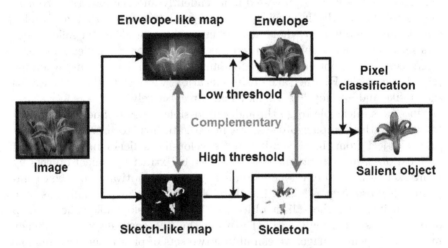

Fig. 4.3 System framework of the approach in (Yu et al, 2010)

When computing **FSM**, a Difference of Gaussian (DoG) filter is used for band pass filtering. The saliency value for any pixel x is computed as:

$$\mathbf{FSM}_x = \|\mathbf{v}_\mu - \overline{\mathbf{v}}_x\|, \tag{4.5}$$

where $\mathbf{v}_\mu$ is the arithmetic mean pixel value of the image and $\overline{\mathbf{v}}_x$ is the feature vector at pixel x from the Gaussian blurred image. Here the pixel feature is chosen as the Lab color. After the computation, the saliency value in **FSM** is normalized to $[0, 1]$.

When computing **CCM**, the center-surround contrast feature in (Liu et al, 2007b) is used by empirically setting several rectangular templates to match the object region and to represent the strip that surrounds the object. As shown in Fig. 4.4, two regions with equal size are selected for each pixel of the image, including a center rectangle and a surrounding strip (as shown in Fig. 4.4). Let $\mathbf{R}_{xy}^c$ and $\mathbf{R}_{xy}^s$ be the RGB color histograms in the center and surround regions, their difference can be computed using the χ^2 distance:

$$\chi^2(\mathbf{R}_x^c, \mathbf{R}_x^s) = \frac{1}{2} \sum_i \frac{(\mathbf{R}_x^c(i) - \mathbf{R}_x^s(i))^2}{\mathbf{R}_x^c(i) + \mathbf{R}_x^s(i)}, \tag{4.6}$$

where $\mathbf{R}_{xy}^c(i)$ and $\mathbf{R}_{xy}^s(i)$ stand for the ith component of the histograms in the center and surround regions, respectively. Compared with other difference measures, such color histograms are insensitive to small changes in size, shape, and viewpoint and can be efficiently computed.

For every pixel x in the image, the width of the center rectangle ranges in $[0.1, 0.7] \times \min(w, h)$, where w and h are image width and height. For each width, five different aspect ratios $\{0.5, 0.75, 1.0, 1.5, 2.0\}$ are tested to find the maximum of the center-surround difference:

$$\{\mathbf{R}_x^{c^*}, \mathbf{R}_x^{s^*}\} = \arg \max_{\mathbf{R}_x^c, \mathbf{R}_x^s} \chi^2(\mathbf{R}_x^c, \mathbf{R}_x^s). \tag{4.7}$$

For each pixel, various center/surround rectangles with different sizes are tested to find the pair of rectangles with the largest difference. Let $T^*(x)$ be the center rectangle that achieves the highest difference, the saliency of x in **CCM** can be computed as:

$$\mathbf{CCM}_x \propto \sum_{x' | x \in T^*(x')} \exp\left(-\frac{d_{x,x'}^2}{\sigma_{x'}^2}\right) \cdot \chi^2(\mathbf{R}_{x'}^{c^*}, \mathbf{R}_{x'}^{s^*}), \tag{4.8}$$

where $d_{x,x'}$ is the distance between pixels x and x'. $\sigma_{x'}^2$ is set to $1/3$ of the size of $T^*(x')$, which is the rectangle centered at x' and contains pixel x. From this definition, the saliency value of a pixel in **CCM** is determined by averaging all the center-surround differences at the pixels that probably belong to the same object (i.e., in the same center rectangle). Finally, the saliency value in **CCM** is also normalized to $[0,1]$.

Center and Surround Regions   Images and center-surround contrast maps

Fig. 4.4 Center-surround regions and region-based contrast maps

Given **FSM** and **CCM**, an envelope-like saliency map $\mathcal{E}$ should be calculated to highlight a rough area that contains the salient object. Moreover, a sketch-like saliency map $\mathcal{S}$ should be calculated to accurately locate part of the salient object. Toward this end, we can calculate $\mathcal{E}$ by simply combine the two feature maps with linear weights:

$$\mathcal{E}_x = \alpha_f \cdot \mathbf{FSM}_x + \alpha_c \cdot \mathbf{CCM}_x, \qquad (4.9)$$

where α_f and α_c are two positive weights to balance **FSM** and **CCM**. Note that we have $\alpha_f + \alpha_c = 1$ to ensure a dynamic range of [0,1].

To construct the sketch-like saliency map, we first roughly estimate the probable center of the salient object. Here we multiply **FSM** and **CCM** to get an integrated saliency map **ISM** in a voting scheme:

$$\mathbf{ISM}_x = \mathbf{FSM}_x^{\beta_f} \cdot \mathbf{CCM}_x^{\beta_c}, \qquad (4.10)$$

where β_f and β_c are two positive constants to balance the influences of **FSM** and **CCM**, respectively. Consequently, the probable center of salient object(s) can be defined as the gravity center of **ISM**, which is denoted as x_g. Given this center, we can build the sketch-like saliency map by highlighting the pixels that form a compact region and distribute near object center.

Following the approach in (Liu et al, 2007b), all colors in the image are represented by Gaussian Mixture Models $\{w_c, \mu_c, \sum_c\}_{c=1}^{C}$, where $\{w_c, \mu_c, \sum_c\}$ is the weight, the mean color and the covariance matrix of the cth component. Each pixel is assigned to a color component with the probability:

$$p(c|x) = \frac{w_c \mathcal{N}(x|\mu_c, \sum_c)}{\sum_{c=1}^{C} w_c \mathcal{N}(x|\mu_c, \sum_c)}. \qquad (4.11)$$

The spatial distribution of the cth Gaussian component can be calculated by the horizontal variance $V_h(c)$ and vertical variance $V_v(c)$. $V_h(c)$ is calculated as:

$$V_h(c) = \frac{\sum_x p(c|x) \cdot |x_h - \bar{x}_h|^2}{\sum_x p(c|x)}, \quad \text{where} \quad \bar{x}_h = \frac{\sum_x p(c|x) \cdot x_h}{\sum_x p(c|x)}, \quad (4.12)$$

where x_h is the x-coordinate of pixel x. The vertical variance $V_v(c)$ can be calculated in a similar way. Consequently, the spatial variance $V(c)$ can be computed as $V(c) = V_x(c) + V_v(c)$.

Beyond the spatial distribution of pixels, we also estimate the pixel distribution of each component against the object center x_g:

$$D(c) = \sum_x p(c|x)d(x, x_g), \quad (4.13)$$

where $d(x, x_g)$ is the distance from pixel x to gravity center x_g. After the calculation, $V(c)$ and $D(c)$ are both normalized to the range $[0, 1]$. Correspondingly, the sketch-like saliency map $\mathcal{S}$ can be estimated as:

$$\mathcal{S}_x \propto \sum_{c=1}^{C} p(c|x)(1 - V(c))(1 - D(c)). \quad (4.14)$$

From (4.14), we see that the sketch-like saliency map will highlight the compact regions near the gravity center of the salient object(s). Recall that the envelope-like saliency map will highlight a large area that contains the salient object(s), the sketch-like and envelope-like saliency maps can be viewed as "complementary."

Given the complementary maps, we have to select a set of pixels to form envelope $\mathbf{E}$ and sketch $\mathbf{S}$. Here we select $\mathbf{E}$ with an adaptive threshold on each image I:

$$\mathbf{E} = \{x | x \in I, \ \mathcal{E}_x \geq \alpha_{low}\frac{\sum_{x \in I} \mathcal{E}_x}{|I|}\}, \quad (4.15)$$

where $|I|$ is the number of pixels in image I. α_{low} is set to a small value to obtain high recall. Moreover, the sketch $\mathbf{S}$ is selected as a subset of the envelope $\mathbf{E}$:

$$\mathbf{S} = \{x | x \in \mathbf{E}, \ \mathcal{S}_x \geq \alpha_{high}\frac{\sum_{x \in \mathbf{E}} \mathcal{S}_x}{|\mathbf{E}|}\}, \quad (4.16)$$

where $|\mathbf{E}|$ is the number of pixels in envelope $\mathbf{E}$. α_{high} is set to a large value to obtain high precision. Some representative envelopes and skeletons are shown in the second and the third rows of Fig. 4.5. We can see that the envelope can well cover the entire salient object while the sketch can accurately reveal part of the salient object.

Finally, the envelope and skeleton can be used as prior knowledge to extract the salient object. Generally speaking, pixels that are not included in the envelope have high probabilities of belonging to the background. Hence, we label them as background seeds. Meanwhile, pixels which are included in the skeleton are likely to belong to the object and we label them as foreground seeds. Then according to the color similarity with the two sets of seeds, we classify the rest pixels in the image into foreground or background. The whole process can be summarized as three major steps:

1. **Seeds clustering**. A background KD-tree and a foreground KD-tree are established for the background and foreground seeds, respectively. Every node in these trees is a cluster of pixel colors.
2. **Pixel assignment**. For every remaining pixel, find the nearest tree node in the color space and then assign the pixel to the cluster. Pixels in foreground clusters are then extracted to form the salient object.
3. **Post-processing**: Isolated components are connected or smoothed to optimize the extracted salient object.

Some representative examples of the segmented salient objects are shown in the last row of Fig. 4.5. It can be seen that the integrity and exactness of extracted objects are very impressive, even when salient objects have similar colors with the background or have unclear boundaries. Generally speaking, the success of this approach is mainly due to the integration of complementary saliency maps. By modeling the reliable foreground and background pixels and classifying the others, the entire salient objects can be effectively detected. However, like most other approaches, this approach may fail when the contrast between salient object and background region is weak, especially when the scenes are complex.

Fig. 4.5 Representative examples for segmenting salient objects using complementary saliency maps. 1st row: images; 2nd row: envelopes; 3rd row: skeletons; 4th row: salient objects. © (2010) Association for Computing Machinery, Inc. Reprinted by permission from (Yu et al, 2010).

4.3   Object-Based Salient Region Detection

Beyond extracting salient objects from location-based saliency maps, an alternative solution for detecting salient objects is to directly measure visual saliency on object-level. That is, the input scene is divided into objects using image segmentation algorithms and these objects will be treated as the basic units when measuring visual saliency. Similar to the location-based saliency models, a widely accepted hypothesis is that visual rarity can also work as a good criterion to quantize object saliency. Following this hypothesis, Liu et al (2007a) first divided images into objects with image segmentation algorithm. After that, the saliency of object was computed by considering its position, size and visual characteristics. Yu and Wong (2007) proposed a real-time clustering approach to segment the input scene into multiple candidate objects and then detect salient objects among them. Aziz and Mertsching (2008) first segmented input scenes into objects by using clustering algorithms in the early stage. Color contrast, size, symmetry, orientation and eccentricity features were then used simultaneously to build five object-based feature maps. These feature maps were finally combined to obtain object saliency. To sum up, the main characteristic of object-based saliency computation is to segment images into objects and use them as the basic units in computing visual saliency.

In early studies, it is often expected that various objects can be perfectly segmented from the input image using fine-tuned object segmentation algorithms. However, it is still a challenging task to segment a complex object as a whole since it main contains multiple parts with different appearances. Yet another problem is that the sizes of various objects may change greatly and it is hard to quantize the rarity by simply measuring the visual difference. To address these two problems, recent studies often aim to segment images into superpixels which are small image patches aligned with intensity edges. Figure 4.6 demonstrates several examples of the superpixels obtained by the segmentation algorithm in (Achanta et al, 2012). From Fig. 4.6, we can see that these superpixels have comparable sizes and perform better than rectangular macro-blocks in preserving object edges. Thus they can be effective in computing object-based saliency maps as well as extracting salient objects. For the sake of simplification, the approaches that segment images into objects with changing sizes are referred to as *region-based* approaches, while the approaches that segment images into objects with comparable sizes are denoted as *superpixel-based* approaches.

In the rest of this Section, we will introduce two region-based approaches in Sects. 4.3.1 and 4.3.2 and two superpixel-based approaches in Sects. 4.3.3 and 4.3.4. Generally speaking, these approaches adopt three main steps to detect salient objects, including: 1) characterizing each region/superpixel using features such as visual appearance and spatial location; 2) generating an object-based saliency map by measuring the irregularity of each region/superpixel

Fig. 4.6 Superpixels are small image patches aligned with intensity edges

and 3) extracting salient objects from the object-based saliency maps. Following these steps, we will introduce these approaches one by one.

4.3.1 Regional Contrasts

Humans pay more attention to image regions that contrast strongly with their surroundings. Moreover, spatial relationships between image regions also play an important role in human attention. High contrast to its surrounding regions is usually stronger evidence for saliency of a region than high contrast to far-away regions. Inspired by this idea, a region contrast analysis approach was proposed in (Cheng et al, 2011) so as to integrate spatial relationships into region-level contrast computation.

In this approach, the input image is first segmented into regions using a graph-based image segmentation method (Felzenszwalb and Huttenlocher, 2004). One benefit to define saliency upon region is related to efficiency. Since the number of regions in an image is far smaller than the number of pixels, computing saliency at region level can significantly reduce the computational cost while producing full-resolution saliency map.

Let r_k be the kth region, it can be represented by the color histogram in Lab color space. Consequently, its saliency value can be computed by measuring its color contrast to all other regions in the image:

$$s(r_k) = \sum_{r_k \neq r_i} w(r_i) D_s(r_k, r_i) D_r(r_k, r_i), \tag{4.17}$$

where $w(r_i)$ is the weight of region r_i. In this approach, $w(r_i)$ is selected as the number of pixels in region r_i since the color contrast to a large region is more meaningful than that to a small region. $D_r(\cdot)$ is the distance metric between the color histograms of the two regions. $D_s(\cdot)$ is a parameter to

re-weight the regional contrast according to their distance, which can be
defined as:

$$D_s(r_k, r_i) = \exp\left(-\frac{d_{ki}}{\sigma_s^2}\right), \tag{4.18}$$

where d_{ki} is the spatial distance between regions r_k and r_i. σ_s controls the
strength of spatial weighting. Larger values of σ_s reduce the effect of spatial
weighting so contrast to farther regions would contribute more to the saliency
of the current region. The spatial distance between two regions is defined as
the Euclidean distance between the centroids of the respective regions. In the
implementation, σ_s^2 is set to 0.4 with pixel coordinates normalized to $[0, 1]$.

Moreover, the color difference between two regions r_1 and r_2 is defined as:

$$D_r(r_1, r_2) = \sum_{i=1}^{n_1}\sum_{j=1}^{n_2} f(c_{1,i})f(c_{2,j})D(c_{1,i}, c_{2,j}), \tag{4.19}$$

where $f(c_{k,i})$ is the frequency of the ith color $c_{k,i}$ among all n_k colors in the
kth region r_k with $k = 1, 2$. $D(c_{1,i}, c_{2,j})$ is the distance between two colors
$c_{1,i}$ and $c_{2,j}$. Note that the frequency of a color occurring in the region is
used as the weight for this color to reflect more the color differences between
dominant colors.

To sum up, the regional saliency can be defined according to the one-vs-all
regional contrasts while taking the influences of region sizes and region distances into account. In Fig. 4.7 we demonstrate several representative saliency
maps computed by this approach. We can see that the performance is tightly
correlated with the image segmentation results. Actually, the main difference between object-based and location-based saliency computation lies in
the changing sizes of the basic saliency units. The influence of object size
must be considered when computing regional saliency. In some cases, trivial image regions may act as "noise" in the result saliency maps, leading to
unsatisfactory results in extracting salient objects.

To address this problem, there may exist two feasible solutions. First, Jiang
et al (2011) adopted the same image segmentation algorithm and proposed
to improve the performance of salient object extraction by incorporating
shape priors. In this manner, the salient trivial regions can be discarded
after saliency computation. Second, Perazzi et al (2012) and Li et al (2013)
proposed to segment image into superpixels aligned with intensity edges with
comparable sizes. In these approaches, the influence of region size is greatly removed before saliency computation. The technical details of these approaches
will be discussed in the rest of this Section.

Fig. 4.7 Saliency maps and salient objects obtained by (Cheng et al, 2011)

4.3.2 Context and Shape Priors

To improve the extraction of large salient objects, Jiang et al (2011) proposed an approach to detect salient objects by incorporating the context and shape priors. By observing the characteristics of salient targets, they introduce three principles to define a salient object:

1. A salient object is always different from its surrounding context.
2. A salient object is most probably placed near image center.
3. A salient object has a well-defined closed boundary.

Note that the second rule is a strong location prior, known as Rule of Thirds. The rule indicates that to attract people's attention, the object of interest, or main element in a photograph should lie at one of the four intersections to approximate the "golden ratio"(about 0.618). For photos with limited objects which are often taken by experienced photographers, this rule works fine to segment the obvious foreground object. However, when encountering complex natural scenes, there may exist numerous candidates of salient objects and these objects may appear far from image centers. In this case, the location assumption may fail, leading to unexpected results.

Following the three principles, regional saliency maps and object shape priors are first extracted for further segmentation. First, images are divided into regions by using the same segmentation algorithm as in (Cheng et al, 2011). However, they adopt a different method to detect a salient object on multiple scales, which is obtained by fragmenting the image with N groups of different parameters. On the nth scale, the input image is segmented into multiple regions $\{r_i^n\}_{i=1}^{R(n)}$. Given a region r_i^n and its spatial neighbors $\{r_k^n\}_{k=1}^{K(n)}$, the saliency of r_i^n can be defined as:

$$s(r_i^n) = -w_i^n \log \left(1 - \sum_{k=1}^{K(n)} \alpha_{ik}^n d_{color}(r_i^n, r_k^n) \right), \tag{4.20}$$

where α_{ik}^n is the ratio between the area of r_k^n and total area of $\{r_k^n\}_{k=1}^{K(n)}$. $d_{color}(r_i^n, r_k^n)$ is the color distance between regions r_i^n and r_k^n, computed as χ^2 distance between the Lab color and hue histograms of the two regions. w_i^n is a center-bias re-weighting factor to emphasize objects around image centers:

$$w_i^n = \exp \left(-9 \left(\left(\frac{dx_i^n}{w} \right)^2 + \left(\frac{dy_i^n}{h} \right)^2 \right) \right), \tag{4.21}$$

where w, h are the width and height of the image, respectively. (dx_i^n, dy_i^n) are the average spatial distance of all pixels in r_i^n to image center.

Note that the image is segmented at N scales and for each scale a region will be assigned a saliency value. Therefore, the saliency value of a pixel p can be derived by averaging the regional saliency:

$$s_m(p) = \frac{1}{Z_p} \sum_{n=1}^{N} \sum_{i=1}^{R(n)} \sum_{p \in r_i^n} \frac{s(r_i^n)}{\|I_p - c_i^n\| + \epsilon}, \tag{4.22}$$

where $Z_p = \sum_{n=1}^{N} \sum_{i=1}^{R(n)} \sum_{p \in r_i^n} (\|I_p - c_i^n\| + \epsilon)^{-1}$ is a normalization factor. c_i^n is the color center of region r_i^n and $\|I_p - c_i^n\|$ is the color distance from the pixel p to the color center. $\epsilon = 0.1$ is a small constant to avoid division by zero.

After calculating the saliency value for each pixel, another concern is to combine them into salient objects, which is similar to salient object detection from location-based saliency maps. Toward this end, an edge detector is used on the image to extract a set of line segments, denoted as $\mathbb{E}$. Then the shape prior, which is an optimal closed contour C^* that covers the salient object, can be achieved by identifying a subset of detected segments in $\mathbb{E}$ and connecting them together. Since many of the detected line segments are disjoint, additional line segments are created to fill the gaps between the endpoints from different detected segments. In this way, the optimal closed contour C^* can be defined as:

$$C^* = \arg \min_C \frac{|C_G|}{\sum_{p \in C} s_m(p)}, \tag{4.23}$$

where $|C_G|$ is the total length of gaps along the contour C, and $\sum_{p \in C} s_m(p)$ is the total saliency value from pixels located inside C. That is, the best closed contour C^* is expected to have the minimum gap length and the maximum saliency inside. Note that an object may consist of several parts, thus multiple optimal contours can be iteratively detected. In each iteration, the saliency values of pixels in existing detected contours are set to 0. These optimal contours are finally combined to form a unique optimal contour.

The optimal contour is actually a polygon with a set of line segments. For pixel-wise segmentation, the shape prior is further represented by a map s_p:

$$s_p(p) = 1 - \exp\left(1 - \gamma d(p)\right), \tag{4.24}$$

where $d(p)$ is the spatial distance between pixel p and optimal closed contour C^*. γ is a confidence of the shape prior, which can be set to 1 in the implementation. From (4.24), we can see that the map s_p will have weak responses only at the pixels around the optimal contour.

Given input image I, saliency map s_m, and shape prior s_p, the problem of salient object detection can be formulated as finding the label set L where $l(p) \in \{0, 1\}$ for each pixel p with 0 for background and 1 for salient object. L can be optimized by solving the following energy minimization problem:

$$L^* = \arg\min_L \sum_{p \in \mathcal{P}} U(p, l_p, s_m) + \lambda \sum_{(p,q) \in \mathcal{N}, l_p \neq l_q} V(p, q, I, s_p), \tag{4.25}$$

where $\mathcal{P}$ is the set of image pixels and $\mathcal{N}$ is a 4-connected neighbor system. The term $U(\cdot)$ is defined as:

$$U(p, l_p, s_m) = \begin{cases} s_m(p) & , \text{if } l_p = 0 \\ 1 - s_m(p) & , \text{if } l_p = 1. \end{cases} \tag{4.26}$$

We can see that minimizing $U(\cdot)$ means selecting pixels with saliency value above 0.5 as foreground. Moreover, the smoothness term $V(\cdot)$ is defined as:

$$V(p, q, I, s_p) = \alpha \cdot \exp\left(-\frac{\|I_p - I_q\|^2}{2\beta}\right) + (1 - \alpha) \cdot \frac{s_p(p) + s_p(q)}{2}, \tag{4.27}$$

where α controls the relative importance of the two terms ($\alpha = 0.5$ in this study). $\beta = \mathbf{E}(\|I_p - I_q\|^2)$. Intuitively, this term encourages the segmentation boundary to be aligned with computed closed contour. Such energy can be efficiently minimized by using the min-cut/max-flow algorithms proposed, leading to a binary segmentation of the image.

The initial saliency map and shape prior are only rough estimation of the salient object. After binary segmentation, both of them can be re-estimated more accurately. Therefore, the salient object can be refined by iteratively updating the saliency map and detecting the salient object. For more details about the iterative optimization, please refer to (Jiang et al, 2011).

In Fig. 4.8, we demonstrate several representative saliency maps and detected salient objects. We can see that by incorporating the shape prior, the result saliency map is much cleaner than those in (Cheng et al, 2011) and the salient objects are more compact. However, unexpected noise may arise when the image segmentation algorithm fails (e.g., the result in the third column). Thus it is necessary to explore other solutions for image segmentation (e.g., superpixel).

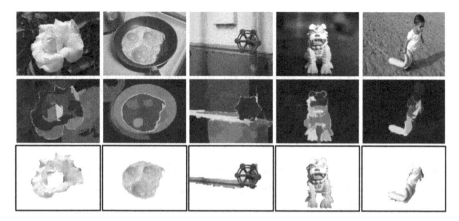

Fig. 4.8 Saliency maps and salient objects obtained by (Jiang et al, 2011)

4.3.3 Saliency Filters

Instead of segmenting images into regions whose sizes may vary remarkably, Perazzi et al (2012) proposed to divide images into superpixels with comparable sizes. Salient objects are then extracted from the superpixel-based saliency maps, which can be computed in four main steps:

1. **Image Segmentation**. In this step, image is over-segmented into superpixels which are perceptually homogeneous regions. In the segmentation, strong contours and edges are preserved and constraints on shape and size should allow for compact, well localized superpixels.
2. **Element Uniqueness**. In this step, contrasts between superpixels are computed to measure their rarity. When using superpixels, variation on the pixel level due to small scale textures or noise is rendered irrelevant, while discontinuities such as strong edges stay sharply localized.
3. **Element distribution**. Beyond the rarity, the spatial variance of color is computed to emphasize compact foreground objects. Approaches based on larger-scale image segmentation (e.g., Cheng et al, 2011) lose this important source of information.
4. **Object Segmentation**. The element uniqueness and distribution properties are jointly considered to measure pixel-accurate saliency map. Salient objects are then extracted from this map using the intelligent threshold as in (4.4).

In their implementation, the image segmentation algorithm proposed in (Achanta et al, 2012) is used in the Lab color space to over-segment images into superpixels. Based on these superpixels, the uniqueness can be defined as the rarity of a segment i given its position $\mathbf{p}_i$ and Lab color $\mathbf{c}_i$ compared to all other segments j:

$$U_i = \sum_{j=1}^{N} \|\mathbf{c}_i - \mathbf{c}_j\|^2 \cdot w(\mathbf{p}_i, \mathbf{p}_j), \tag{4.28}$$

where $w(\mathbf{p}_i, \mathbf{p}_j)$ is used to re-weight the influence between two superpixels with respect to their distance. In this approach, it is defined as Gaussian weight:

$$w(\mathbf{p}_i, \mathbf{p}_j) = \frac{1}{Z_i} \exp\left(-\frac{\|\mathbf{p}_i - \mathbf{p}_j\|^2}{2\sigma_p^2}\right), \tag{4.29}$$

where σ_p is set to 0.25 for a balance between local and global effects. Z_i is a normalization factor to ensure that $\sum_{j=1}^{N} w(\mathbf{p}_i, \mathbf{p}_j) = 1$. Consequently, (4.28) can be decomposed as:

$$U_i = \mathbf{c}_i^2 - 2\mathbf{c}_i \sum_{j=1}^{N} \mathbf{c}_j w(\mathbf{p}_i, \mathbf{p}_j) + \sum_{j=1}^{N} \mathbf{c}_j^2 w(\mathbf{p}_i, \mathbf{p}_j), \tag{4.30}$$

where the second term the third term can be viewed as using a Gaussian blurring kernel on color $\mathbf{c}_j$ and the squared color $\mathbf{c}_j^2$. In this manner, the uniqueness can be computed in linear time without crude approximations such as histograms or distance to mean color.

Consequently, the element distribution for a segment i can be measured using the spatial variance D_i of its color $\mathbf{c}_i$:

$$D_i = \sum_{j=1}^{N} \|\mathbf{p}_j - \mu_i\|^2 w(\mathbf{c}_i, \mathbf{c}_j), \tag{4.31}$$

where $w(\mathbf{c}_i, \mathbf{c}_j)$ is used to describe the color similarity of color $\mathbf{c}_i$ and color $\mathbf{c}_j$. $\mu_i = \sum_{j=1}^{N} w(\mathbf{c}_i, \mathbf{c}_j)\mathbf{p}_j$ is the weighted mean position of color $\mathbf{c}_i$. Similarly, the color similarity can be defined as Gaussian term:

$$w(\mathbf{c}_i, \mathbf{c}_j) = \frac{1}{Z_i} \exp\left(-\frac{\|\mathbf{c}_i - \mathbf{c}_j\|^2}{2\sigma_c^2}\right), \tag{4.32}$$

where σ_c is set to 20 to control the sensitivity in measuring color similarity. Z_i is a normalization factor ensuring $\sum_{j=1}^{N} w(\mathbf{c}_i, \mathbf{c}_j) = 1$. Similar to (4.30), the spatial variance can be also evaluated in linear time by factoring out:

$$D_i = \sum_{j=1}^{N} \mathbf{p}_j^2 w(\mathbf{c}_i, \mathbf{c}_j) - \left(\sum_{j=1}^{N} \mathbf{p}_j w(\mathbf{c}_i, \mathbf{c}_j)\right)^2, \tag{4.33}$$

where the first term and the second term can be viewed as blurring the squared position $\mathbf{p}_j^2$ and position $\mathbf{p}_j$ in the 3-dimensional color space.

After the computation, the uniqueness and spatial variance are then normalized to the dynamic range of $[0,1]$. Thus the saliency value for a superpixel can be computed as:

$$s_i = U_i \cdot \exp(-k \cdot D_i), \tag{4.34}$$

where $k = 6$ is a parameter to emphasize the contribution of D_i. From (4.34), we can see that a superpixel which is rare (i.e., large U_i) and compact (i.e., small D_i) will become salient.

To get the full resolution saliency maps, the saliency value $\widetilde{s}_i$ of a pixel can be derived as a weighted linear combination of the saliency from its surrounding superpixels:

$$\widetilde{s}_i = \sum_{j=1}^{N} \frac{1}{Z_i} \exp\left(-\frac{\alpha\|\mathbf{c}_i - \mathbf{c}_j\|^2 + \beta\|\mathbf{p}_i - \mathbf{p}_j\|^2}{2} \right) s_j, \tag{4.35}$$

where α and β are parameters controlling the sensitivity to color and position. In the experiment, α and β are both set to $1/30$.

In Fig. 4.9, we demonstrate several representative saliency maps and detected salient objects. We can see that assigning pixel saliency by combining all the superpixel saliency values can generate impressive viewing effect, which is a highlight of this approach. With these saliency maps, a simple thresholding operation can generate impressive results in salient object segmentation. Moreover, all involved operations are formulated within a single high-dimensional Gaussian filtering framework, making the algorithm extremely efficient.

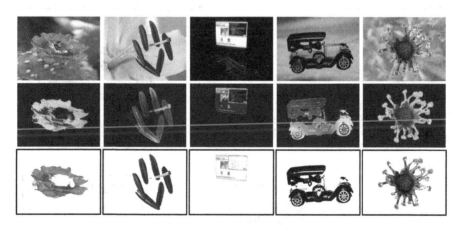

Fig. 4.9 Saliency maps and salient objects obtained by (Perazzi et al, 2012)

4.3.4 Single Image Optimization

Usually, the approaches discussed above can achieve promising performance using predefined or learned saliency measurement criteria. However, one drawback of these approaches is that they often adopt unified models to process various images, while these saliency hypotheses (e.g., rarity, center-bias and correlation hypotheses) may not always hold in different images. For instance, using the rarity hypothesis can easily detect small salient objects but may fail when processing images with large salient objects. Therefore, various saliency hypotheses should be adaptively taken into account to obtain the best saliency map on each specific image.

Toward this end, an approach is proposed in (Li et al, 2013) to estimate visual saliency through single image optimization. Instead of mapping visual features to saliency values with a unified model, the saliency values of all regions are treated as the optimization objective on each single image. In this process, the saliency value of each region is optimized when considering the influences of all the other image regions. By using a quadratic programming framework, these saliency values can be adaptively optimized on each image to simultaneously meet several saliency hypotheses on visual rarity, center-bias and mutual correlation.

To estimate visual saliency, an image I is first segmented into N regions using the algorithm (Achanta et al, 2012), denoted as $\{\mathcal{B}_i\}_{i=1}^N$. As shown in Fig. 4.10, these compact regions can well preserve object shapes and are much easier to obtain than the perfectly segmented objects. Moreover, such compact segmentation avoid the ambiguities around object boundaries which often arise when simply partitioning images into macro blocks with fixed sizes.

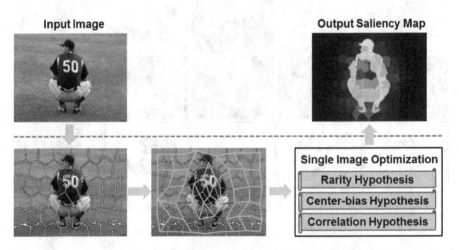

Fig. 4.10 System framework of the approach in (Li et al, 2013). Copyright © IEEE. All rights reserved. Reprinted, with permission, from (Li et al, 2013).

Given $\{\mathcal{B}_i\}_{i=1}^N$, visual saliency can be computed as a kind of *regional rarity*. That is, high saliency values should be assigned to unique or rare regions. To quantize such rarity, the mutual correlations between all regions should be derived. Inspired by the idea that visually similar regions should have strong correlations, $\mathcal{B}_i$ can be represented by its visual appearance $\mathbf{v}_i$, which is a column vector computed by averaging all the pixel-wise intensity, red-green opponency and blue-yellow opponency in $\mathcal{B}_i$. Here the approach in (Walther and Koch, 2006) is used to compute the red-green opponency RG_v and blue-yellow opponency BY_v for pixel v:

$$RG_v = \frac{r_v - g_v}{\max(r_v, g_v, b_v)},$$
$$BY_v = \frac{b_v - \min(r_v, g_v)}{\max(r_v, g_v, b_v)}, \tag{4.36}$$

where r_v, g_v, b_v are the red, green and blue values of pixel v. Similar to (Walther and Koch, 2006), we set $RG_v = BY_v = 0$ if $\max(r_v, g_v, b_v) < 0.1$ to avoid large fluctuations at low luminance.

After extracting $\{\mathbf{v}_i\}_{i=1}^N$, the correlation w_{ij} between $\mathcal{B}_i$ and $\mathcal{B}_j$ can be computed as a kind of visual similarity:

$$w_{ij} = \exp\left(-\frac{\|\mathbf{v}_i - \mathbf{v}_j\|_1}{3}\right). \tag{4.37}$$

From the definition in (4.37), we can see that the mutual correlation between any two image regions is symmetric (i.e., $w_{ij} = w_{ji}$). Note that the spatial distance between $\mathcal{B}_i$ and $\mathcal{B}_j$ is not considered in (4.37) to avoid the boundary effect. Usually, severe boundary effect may arise when incorporating the influence of such spatial distance. In this case, regions far from image centers will have relatively weak correlations with all the other regions. Consequently, high saliency values could be mistakenly assigned to the regions around image corners.

After measuring the mutual correlations, the saliency value for each region can be computed from such correlations. Instead of using a model that directly maps regional visual features to saliency values, the saliency values of all regions in a specific image are simultaneously optimized to best meet several saliency hypotheses, including:

- **Rarity hypothesis**: an image region, which only has weak correlations (i.e., low visual similarities) with all the other regions, should probably be salient;
- **Center-bias hypothesis**: an image region, which appears near to image center, should probably be salient.
- **Correlation hypothesis**: two tightly correlated image regions should have similar saliency values if they are near to each other;

Following these hypotheses, the problem of visual saliency estimation is formulated into an optimization framework. In this framework, saliency values of various regions can be optimized simultaneously to best meet the hypotheses. Let s_i be the saliency value of $\mathcal{B}_i$, $\{s_i\}_{i=1}^{N}$ can be optimized by solving:

$$\min_{\{s_i\}_{i=1}^{N}} \sum_{i=1}^{N} s_i \sum_{j \neq i}^{N} w_{ij} + \lambda_c \sum_{i=1}^{N} s_i e^{d_i/d_D}$$

$$+ \lambda_r \sum_{i=1}^{N} \sum_{j \neq i}^{N} (s_i - s_j)^2 w_{ij} e^{-d_{ij}/d_D} \qquad (4.38)$$

$$s.t. \ \ 0 \leq s_i \leq 1, \forall i,$$

$$\sum_{i=1}^{N} s_i = 1.$$

where d_D is half the image diagonal length. d_{ij} and d_i are the distances from $\mathcal{B}_i$ to $\mathcal{B}_j$ and image center, respectively. λ_c and λ_r are two weights to balance the influences of these terms, which can be automatically estimated as:

$$\lambda_c = \sqrt{N}, \ \ \lambda_r = \log N. \qquad (4.39)$$

As shown in (4.38), high penalties will arise if: 1) assigning high saliency to a region that is tightly related to all the other regions(the first term); 2) allocating high saliency to regions near to image boundaries (the second term) or 3) assigning different saliency values to tightly related regions that are near to each other (the third term). Note that the optimization problem in (4.38) only has quadratic and linear terms with linear constraints, thus it can be solved with the active-set algorithm by iteratively searching for the active constraints and solving the equality problems with Lagrange multipliers.

After computing the region-based saliency values, another problem is how to extract the salient objects. As shown in Fig. 4.11, two intelligent thresholds are adopted to extract the salient objects. First, these two thresholds are computed to select the most reliable foreground and background regions:

$$T_{low} = \frac{1}{N} \sum_{i=1}^{N} s_i, \ T_{high} = \frac{2}{N} \sum_{i=1}^{N} s_i. \qquad (4.40)$$

After that, regions with saliency values higher than T_{high} and lower than T_{low} are selected as reliable foreground and background regions, respectively. In particular, the top 5% regions are directly selected as reliable foreground regions if $T_{high} > \max\{s_i\}_{i=1}^{N}$.

Given the reliable foreground and background regions, other regions are then classified according to their mutual correlations, which is similar to the approach in (Yu et al, 2010). Let $\mathbb{I}_{high}$ and $\mathbb{I}_{low}$ be the indices of the reliable

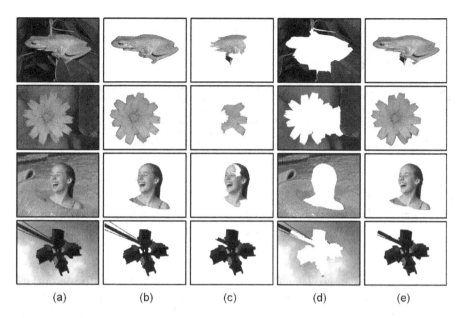

foreground and background regions, $\mathcal{B}_i$ with saliency value $T_{low} \leq s_i \leq T_{high}$ will be classified as a foreground region if:

$$\max\{w_{ij}, j \in \mathbb{I}_{high}\} > \max\{w_{ij}, j \in \mathbb{I}_{low}\}. \tag{4.41}$$

Otherwise, $\mathcal{B}_i$ will be classified as a background region. Different from (Achanta et al, 2009) and (Perazzi et al, 2012) which directly binarize saliency maps using one threshold like T_{high}, the reliable background regions are also selected using T_{low} and such background regions can help to recover the foreground regions whose saliency values are in $[T_{low}, T_{high}]$. In this manner, most salient regions can pop-out after the binarization, especially when the salient objects are very large (as shown in Fig. 4.11).

Some representative examples generated by this approach are illustrated Fig. 4.12. We can see that the approach performs impressive on detecting large salient objects. Actually, the main advantage of this approach is that it can *simultaneously optimizes* the saliency values of all regions in a specific image. In the optimization, the saliency value of a specific region is computed when the influences of all the other regions are taken into account. In this manner, the estimated saliency map can adaptively meet all the three saliency hypotheses on visual rarity, center-bias and mutual correlation.

Fig. 4.12 Saliency maps and salient object obtained by (Li et al, 2013)

Moreover, we can see that the framework is flexible. Given specific prior knowledge on the salient targets (e.g., the salient targets that may appear in the image), new penalty terms can be simply added into (4.38) to bring in such prior knowledge to improve the estimated saliency maps. Moreover, the estimated saliency maps have the same size as the input image. As shown in Fig. 4.12, the boundaries of the salient objects can be well maintained. This is an useful characteristic since many object analysis techniques need to extract features from boundaries.

4.4   Model Comparison

In this Section, we conduct an experiment to validate the effectiveness of all the object-based saliency models discussed above. In the experiment, we adopt the image saliency benchmark proposed in (Achanta et al, 2009). This benchmark, denoted as **ASD**, contains 1,000 images with obvious salient objects. In each image, salient objects are labeled with accurate masks. This benchmark has been used by many studies for evaluation.

On this benchmark, 14 visual saliency models are compared. These approaches can be roughly categorized into three groups, including:

- **Location-based group**: This group contains 8 approaches that detect *salient locations* in images, including (Itti et al, 1998; Bruce and Tsotsos, 2005; Harel et al, 2007; Hou and Zhang, 2007, 2009; Goferman et al, 2010; Wang et al, 2010; Riche et al, 2012);
- **Object-based group**: this group contains 6 approaches that detect *salient objects* from images, including (Achanta et al, 2009; Yu et al, 2010; Cheng et al, 2011; Jiang et al, 2011; Perazzi et al, 2012; Li et al, 2013).

In the comparison, we use **Recall**, **Precision** and $\mathbf{F}_\beta$ to evaluate the performance of extracting salient objects. In the comparison, all the other saliency maps are binarized using the intelligent thresholds computed by (4.4). This threshold is also used in some other approaches such as (Perazzi et al, 2012; Achanta et al, 2009). Moreover, we compute $\mathbf{F}_\beta$ by equally considering **Recall** and **Precision**:

$$\mathbf{F}_\beta = \frac{2 \times \mathbf{Recall} \times \mathbf{Precision}}{\mathbf{Recall} + \mathbf{Precision}}. \tag{4.42}$$

Table 4.1 Performance of various approaches on **ASD**

	Algorithm	Recall	Precision	$\mathbf{F}_\beta$
	Itti et al (1998)	0.29	0.69	0.41
	Bruce and Tsotsos (2005)	0.37	0.53	0.44
	Harel et al (2007)	0.38	0.67	0.49
Location-based	Hou and Zhang (2007)	0.36	0.50	0.42
Saliency Model	Hou and Zhang (2009)	0.41	0.60	0.49
	Goferman et al (2010)	0.47	0.62	0.54
	Wang et al (2010)	0.50	0.63	0.56
	Riche et al (2012)	0.53	0.68	0.59
	Vikram et al (2012)	0.46	0.70	0.56
	Achanta et al (2009)	0.51	0.76	0.61
Object-based	Yu et al (2010)	0.89	0.88	0.89
Saliency Model	Cheng et al (2011)	0.47	0.87	0.61
	Jiang et al (2011)	0.76	0.89	0.82
	Perazzi et al (2012)	0.71	0.90	0.79
	Li et al (2013)	0.81	0.86	0.83

From Table 4.1, we can see that the object-based saliency models outperform almost all the location-based saliency models in **Precision**. This is reasonable since the location-based models aim to pop-out all the probable salient targets, while the object-based models focus on detecting only the most salient target as a whole. The assumption on the number of salient objects is a strong prior since most of the 1,000 images contain only one salient object. Actually, when processing natural images with complex content (e.g., images in the benchmark proposed by Judd et al, 2009), the location-based approaches often outperforms the object-based approaches.

Moreover, among all the six object-based saliency models, we can see that (Achanta et al, 2009; Cheng et al, 2011; Jiang et al, 2011; Perazzi et al, 2012) have much higher precision than recall. Thus they argue that **Precision** is much more important than **Recall** and emphasize only **Precision** when compute the $\mathbf{F}_\beta$. However, we can see that the **Recall** of these approaches are not very promising and it is not so easy to reach a high **Recall** while maintaining high **Precision**. Actually, some applications such as mobile search and video retargeting, **Recall** is as important as **Precision** since these applications often rely on the features extracted from object boundaries. When only emphasizing **Precision**, some object parts will be wrongly suppressed and the "fake" boundaries will mislead these applications. Therefore, we equally treat **Recall** and **Precision** instead of emphasizing only **Precision** when calculate $\mathbf{F}_\beta$. In this case, we can see that the approach in (Yu et al, 2010) achieves the best overall performance. However, this approach is very slow since it has to test various center/surround templates for each pixel to compute the location-based saliency. Moreover, there are many parameters in (Yu et al, 2010) that need to be manually fine-tuned, which decreases its generalization ability.

4.5   Notes

In this Chapter we introduce several approaches for object-based saliency computation. These object-based approaches still adopt the same hypothesis used in the location-based approaches. That is, salient visual subsets are unique or rare. However, the main difference is that the basic saliency unit is changed from spatial location to object (i.e., region/superpixel). Consequently, the saliency measurement is changed from local/global contrasts to regional contrasts. Moreover, by emphasizing the compactness, these approaches can detect the salient object as a whole instead of only popping-out the salient borders. In experiments, we can see that these approaches can give impressive results on images with simple background and obvious salient objects. However, they may fail when processing scenes with complex background and multiple foreground objects.

One main drawback of these object-based approaches, as well as the location-based approaches, is that they failed to take the prior knowledge into account. The prior knowledge, which can be learned from other scenes, can help to optimize the saliency detection process in at least three ways. First, the feature prior (i.e., the "what" prior) can reveal the features which can best distinguish salient targets from distractors in specific scenes. Second, the location prior (i.e., the "where" prior) can tell the probable locations that the salient objects may appear, leading to high precision in salient location/object detection. Third, the correlation prior, which encodes the inherent correlations between various locations/regions, can help to pop-out the complex salient object as a whole, even when it consists of multiple parts with distinct appearances. Since the prior knowledge is very important for perfect salient location/object detection, in the next several Chapters we will discuss how to learn the prior knowledge with machine learning algorithms and how to integrate such priors to address the problem of visual saliency computation.

References

Achanta, R., Hemami, S., Estrada, F., Süsstrunk, S.: Frequency-tuned salient region detection. In: Preceedings of the IEEE Conference on Computer Vision and Pattern Recognition (CVPR), pp. 1597–1604 (2009), doi:10.1109/CVPR.2009.5206596

Achanta, R., Shaji, A., Smith, K., Lucchi, A., Fua, P., Süsstrunk, S.: Slic superpixels compared to state-of-the-art superpixel methods. IEEE Transactions on Pattern Analysis and Machine Intelligence 34(11), 2274–2282 (2012), doi:10.1109/TPAMI.2012.120

Aziz, M., Mertsching, B.: Fast and robust generation of feature maps for region-based visual attention. IEEE Transactions on Image Processing 17(5), 633–644 (2008), doi:10.1109/TIP.2008.919365

Bruce, N.D., Tsotsos, J.K.: Saliency based on information maximization. In: Advances in Neural Information Processing Systems (NIPS), Vancouver, BC, Canada, pp. 155–162 (2005)

Cheng, M.M., Zhang, G.X., Mitra, N., Huang, X., Hu, S.M.: Global contrast based salient region detection. In: Preceedings of the IEEE Conference on Computer Vision and Pattern Recognition (CVPR), pp. 409–416 (2011), doi:10.1109/CVPR.2011.5995344

Felzenszwalb, P.F., Huttenlocher, D.P.: Efficient graph-based image segmentation. International Journal of Computer Vision 59(2), 167–181 (2004), doi:10.1023/B:VISI.0000022288.19776.77

Goferman, S., Zelnik-Manor, L., Tal, A.: Context-aware saliency detection. In: Preceedings of the IEEE Conference on Computer Vision and Pattern Recognition (CVPR), pp. 2376–2383 (2010), doi:10.1109/CVPR.2010.5539929

Harel, J., Koch, C., Perona, P.: Graph-based visual saliency. In: Advances in Neural Information Processing Systems (NIPS), pp. 545–552 (2007)

Hou, X., Zhang, L.: Saliency detection: A spectral residual approach. In: Proceedings of the IEEE Conference on Computer Vision and Pattern Recognition (CVPR), pp. 1–8 (2007), doi:10.1109/CVPR.2007.383267

Hou, X., Zhang, L.: Dynamic visual attention: Searching for coding length increments. In: Advances in Neural Information Processing Systems (NIPS), pp. 681–688 (2009)

Itti, L., Koch, C., Niebur, E.: A model of saliency-based visual attention for rapid scene analysis. IEEE Transactions on Pattern Analysis and Machine Intelligence 20(11), 1254–1259 (1998), doi:10.1109/34.730558

Jiang, H., Wang, J., Yuan, Z., Liu, T., Zheng, N.: Automatic salient object segmentation based on context and shape prior. In: Proceedings of the British Machine Vision Conference, pp. 110.1–110.12. BMVA Press (2011), doi:10.5244/C.25.110

Judd, T., Ehinger, K., Durand, F., Torralba, A.: Learning to predict where humans look. In: Proceedings of the IEEE International Conference on Computer Vision (ICCV), pp. 2106–2113 (2009), doi:10.1109/ICCV.2009.5459462

Li, J., Tian, Y., Duan, L., Huang, T.: Estimating visual saliency through single image optimization. IEEE Signal Processing Letters 20(9), 845–848 (2013), doi:10.1109/LSP.2013.2268868

Liu, H., Jiang, S., Huang, Q., Xu, C., Gao, W.: Region-based visual attention analysis with its application in image browsing on small displays. In: Proceedings of the 15th Annual ACM International Conference on Multimedia, MULTIMEDIA 2007, pp. 305–308. ACM, New York (2007a), doi:10.1145/1291233.1291298

Liu, T., Sun, J., Zheng, N.N., Tang, X., Shum, H.Y.: Learning to detect a salient object. In: Preceedings of the IEEE Conference on Computer Vision and Pattern Recognition (CVPR), pp. 1–8 (2007b), doi:10.1109/CVPR.2007.383047

Ma, Y.F., Zhang, H.J.: Contrast-based image attention analysis by using fuzzy growing. In: Proceedings of the 11th ACM International Conference on Multimedia, MULTIMEDIA 2003, pp. 374–381. ACM, New York (2003), doi:10.1145/957013.957094

Perazzi, F., Krahenbuhl, P., Pritch, Y., Hornung, A.: Saliency filters: Contrast based filtering for salient region detection. In: Preceedings of the IEEE Conference on Computer Vision and Pattern Recognition (CVPR), pp. 733–740 (2012), doi:10.1109/CVPR.2012.6247743

Riche, N., Mancas, M., Gosselin, B., Dutoit, T.: Rare: A new bottom-up saliency model. In: Preceedings of the 19th IEEE International Conference on Image Processing (ICIP), pp. 641–644 (2012), doi:10.1109/ICIP.2012.6466941

Vikram, T.N., Tscherepanow, M., Wrede, B.: A saliency map based on sampling an image into random rectangular regions of interest. Pattern Recognition, 3114–3124 (2012)

Walther, D., Koch, C.: Modeling attention to salient proto-objects. Neural Networks 19(9), 1395–1407 (2006)

Wang, W., Wang, Y., Huang, Q., Gao, W.: Measuring visual saliency by site entropy rate. In: Proceedings of the IEEE Conference on Computer Vision and Pattern Recognition (CVPR), pp. 2368–2375 (2010), doi:10.1109/CVPR.2010.5539927

Yu, H., Li, J., Tian, Y., Huang, T.: Automatic interesting object extraction from images using complementary saliency maps. In: Proceedings of the International Conference on Multimedia, MULTIMEDIA 2010, pp. 891–894. ACM, New York (2010), doi:10.1145/1873951.1874105

Yu, Z., Wong, H.S.: A rule based technique for extraction of visual attention regions based on real-time clustering. IEEE Transactions on Multimedia 9(4), 766–784 (2007), doi:10.1109/TMM.2007.893351

Chapter 5
Learning-Based Visual Saliency Computation

In this Chapter, we describe how to model the top-down factors in human vision system. Usually, such top-down factors should be learned from user data such as fixations or labeled salient objects by using machine learning algorithms. Therefore, we will present how to model various top-down factors by using supervised or unsupervised learning algorithms. Moreover, we aim to show how the machine learning algorithms can help to improve the performance of saliency models.

5.1 Overview

Generally, determining salient region requires finding the value, location, and extent of the rare subset from the input visual stimuli. Since such stimuli can be viewed as a package of preattentive feature channels, one feasible solution is to detect the rare signals from each channel and the final salient subsets can be obtained by combining the results across channels. In this process, the most significant difference between bottom-up and top-down models is the way to perform feature combination. The combination process in most existing bottom-up models is driven by the intrinsic attributes of the stimuli. That is, the unique or rare subsets from different feature channels are combined without any bias. However, the bottom-up saliency models motivated by visual rarity detection often fail to perfectly pop-out targets and suppress distractors. As shown in Fig. 5.1, certain parts of the salient target may be suppressed by the model when using improper scales, while some distractors may pop-out since they are distinct from their surroundings. To recover such wrongly suppressed targets and inhibit improperly popped-out distractors, the unbiased combination is far from sufficient. Therefore, it is necessary to incorporate the top-down factors such as tasks and prior knowledge to bias the competition.

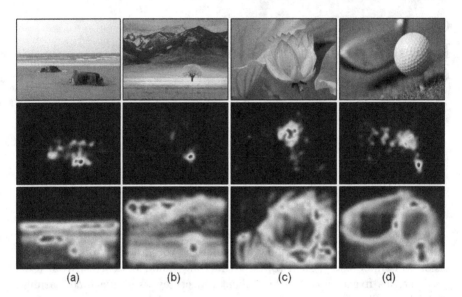

(a)	(b)	(c)	(d)

Fig. 5.1 Examples of wrongly suppressed targets and improperly popped-out distractors. 1st row: input images; 2nd row: fixation density map; 3rd row: saliency maps obtained by (Itti et al, 1998). Reproduced from (Li et al, 2013), with kind permission from Springer Science+Business Media.

In top-down approaches, the feature combination can be modulated by task and/or prior knowledge, demonstrating a biased selectivity on the input stimuli (Mozer et al, 2005). Biological evidence verifies that the neurons linked with various stimuli undergo a mutual competition to generate the bottom-up saliency, while the top-down process can bias such competition in favor of a specific category of stimuli (Frith, 2005). Note that in Chap. 1 we have presented that there are two major types of top-down processes in human brain, including the volitional and mandatory top-down processes. The volitional top-down process involves acts of will and can be viewed as task-relevant in many cases. On the contrary, the mandatory top-down process is considered to be related to prior knowledge derived from past scenes and may work even when the subjective willing is absent. According to the two types of top-down processes, existing top-down approaches can be classified into two categories: the task-driven category and the knowledge-based category. Approaches in the task-driven category mainly consider the influence of volitional top-down process and focus on learning the task-related top-down factors. On the contrary, approaches in the knowledge-based category focus on the mandatory top-down process and aim to learn the experience and prior knowledge under free-viewing conditions. In this Chapter, we will review the approaches in these two categories and describe the technical details of one representative approach from each category.

5.2   Task-Driven Saliency Model

Some of the learning-based approaches treat the top-down factors as predefined tasks and incorporate them to predict visual saliency in task-specific
scenarios. From the neurobiological perspective, such tasks can be viewed as
acts of will and correspond to the *volitional top-down process* in our vision system. For example, Torralba et al (2006) proposed to learn the task-dependent
priors that can be integrated into a Bayesian framework to pop-out the objects corresponding to specific search tasks. In their approach, they explored
the roles of different searching tasks such as people search, painting search
and mug search. Similarly, Chikkerur et al (2010) also proposed a Bayesian
inference theory for visual attention computation. In this theory, attention is
viewed as part of the inference process to solve the visual recognition problem
of what is where. That is, the main goal of the visual system is to infer the
identity and the position of objects in visual scenes: spatial attention emerges
as a strategy to reduce the uncertainty in shape information while feature-
based attention reduces the uncertainty in spatial information. Feature-based
and spatial attention represent two distinct modes of a computational process
solving the problem of recognizing and localizing objects. Using this theory,
they proposed to learn the priors used for searching cars and pedestrians in
images, with which they trained models to attend to task-related targets.

One drawback of these task-driven approaches is that they can only adapt
to specific task-related scenarios (e.g., a scene with cars and pedestrians).
Consequently, such task-related priors, which are often defined on specific
object classes, may prevent their further usage in generic scenarios. Actually,
it is difficult to explicitly predefine such tasks in most real-world scenes.
Therefore, it is necessary to explore how to infer the latent search tasks
people may perform in free-viewing conditions.

From the perspective of task modeling, we propose a probabilistic multi-
task learning approach in this Section. In this approach, the problem of visual
saliency computation is modeled by simultaneously considering the stimulus-
driven and task-related factors in a probabilistic framework. In this framework, the task-related "stimulus-saliency" functions and various model fusion
strategies are learnt from scenes in the training data and then transferred to
new scenes for estimating their visual saliency maps. Since an observer may
not always maintain an explicit task in mind, the visual attention is often
guided by the experience for the past similar scenes to search the expected
targets (Chun, 2005). Therefore, the term "task" particularly refers to such
visual search task in this study. With this definition, we focus on the
following three issues:

1. **Which** tasks are probably adopted for a scene in free-viewing conditions?
2. **What** are the task-related "stimulus-saliency" functions?
3. **How** to adaptively combine the bottom-up and top-down factors in
   different scenes?

As shown in Fig. 5.2, the framework of this approach consists of two main modules: the bottom-up (stimulus-driven) and top-down (task-related) components. In the bottom-up component, preattentive features are first selected from the input scene. Afterwards, the energy maps are computed from these features using multi-scale wavelet decomposition. Consequently, global and local descriptors are extracted to characterize a scene and its blocks. These local descriptors are fused using the feature integration theory and the bottom-up saliency map is then generated through unbiased feature competition. In contrast to the straightforward processing in the bottom-up component, a probabilistic multi-task learning algorithm is first used in the top-down component to learn the typical task-related "stimulus-saliency" functions and fusion strategies from the training data. They are then used as priors for visual saliency computation in similar scenes to provide task-driven predictions. Finally, the bottom-up and top-down predictions are fused in the probabilistic framework to obtain the visual saliency using the fusion strategies adopted in the past similar scenes.

The main contributions of this approach are summarized as follows:

1. A bottom-up visual saliency model is proposed based on multi-scale wavelet decomposition and unbiased feature competition. The model efficiently detects irregular subsets in scenes by simulating the mechanisms of retinal cells using low-pass and high-pass filters. These irregularities are then integrated and competed to output the bottom-up saliency. Experiments on two eye-fixation datasets (Itti, 2008) and one regional saliency dataset (Li et al, 2009) show that this model outperforms seven existing bottom-up approaches.
2. We propose a multi-task learning algorithm to learn the task-related "stimulus-saliency" functions for each scene. The fusion strategies are also learnt simultaneously. In the framework of multi-task learning, incorporating an appropriate sharing of information between tasks and fusion strategies can generate more robust results.
3. We propose a probabilistic framework for visual saliency computation. The bottom-up and top-down factors are adaptively fused in this framework to output the visual saliency. Experimental results verify that this approach remarkably outperforms the pure bottom-up models (e.g., Itti and Baldi, 2005; Guo et al, 2008) and learning-based approaches (e.g., Peters and Itti, 2007).

In the rest part of this Section, we will first describe the bottom-up saliency component of the proposed approach. After that, the probabilistic multi-task learning approach is presented. Finally, extensive experiments will be conducted to validate the effectiveness of the proposed approach.

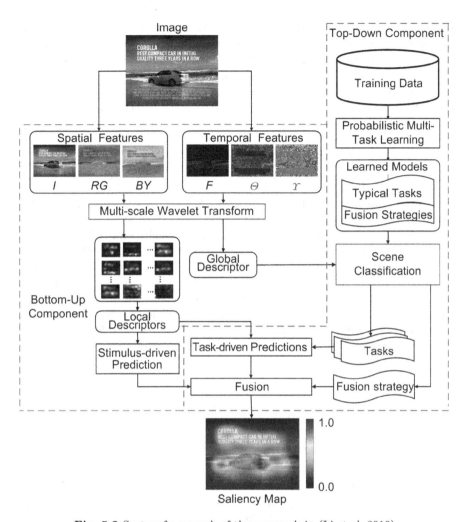

Fig. 5.2 System framework of the approach in (Li et al, 2010)

5.2.1   The Bottom-Up Component

Generally, the bottom-up component should simulate the low-level processes
in human vision system to provide stimulus-driven saliency. Toward this end,
we present a bottom-up model by simulating the filtering processes in retinal
cells to detect the irregular subsets in a scene. The model then simulates
the feature integration and unbiased feature competition processes in human
brain to integrate these irregularities for bottom-up saliency. In the simulation process, an important step is to extract effective scene representations.
Among these representations, local descriptors are inferred from the input

signals to reveal the spatial distribution of the preattentive stimuli, while a global descriptor is extracted to characterize the whole scene for memorization and cognition. In this Section, we will focus on how to select preattentive features from a scene and then extract local and global scene representations from these features. We will also discuss how to infer the bottom-up saliency from the local descriptors through feature integration and competition.

5.2.1.1   Preattentive Features

In general, a video sequence can be expressed as a conjugate of information flows from multiple visual feature channels. Among these channels, some are the probable sources of preattentive guidance (e.g., color, luminance and motion) while the others are likely not. Based on the conclusion of (Wolfe, 2005), we select three probable preattentive features (i.e., luminance, red-green and blue-yellow opponencies) for extracting spatial representations and the other three (i.e., flicker, motion direction and motion strength) for extracting temporal representations. Let r_k, g_k, b_k be the red, green and blue channels of the k-th scene in a video sequence, then we can compute the three spatial preattentive features, i.e., luminance I_k, red-green opponency RG_k and blue-yellow opponency BY_k, according to (Walther and Koch, 2006):

$$I_k = \frac{r_k + g_k + b_k}{3},$$

$$RG_k = \frac{r_k - g_k}{\max(r_k, g_k, b_k)}, \qquad (5.1)$$

$$BY_k = \frac{b_k - \min(r_k, g_k)}{\max(r_k, g_k, b_k)}.$$

Note that all the calculations are carried out at the pixel level. We also compute three preattentive features including flicker F_k, motion direction Θ_k and motion strength Υ_k to depict the temporal scene representations. First, the motion vector is estimated at each pixel through fast block motion computation (Cheung and Po, 2002). Let u_k and v_k be the maps of displacements in the horizontal and vertical directions, the temporal preattentive features can be computed as:

$$F_k = I_k - I_{k-1},$$

$$\Theta_k = \tan^{-1}\left(\frac{v_k}{u_k}\right), \qquad (5.2)$$

$$\Upsilon_k = \sqrt{u_k^2 + v_k^2}.$$

5.2.1.2   Scene Representations

Inspired by the retina model proposed by (Marat et al, 2009), we extract the representations for each location in a scene by simulating the low-level processes in retinal cells. As shown in the retina model in the upper part of Fig. 5.3, an incoming scene is first smoothed by the horizontal cells (corresponding to the first low-pass filtering). After that, the bipolar cells compute the differences between the incoming scene and the smoothed versions to output the high-frequency components (corresponding to the high-pass filtering). Finally, the second smoothing operation is carried out by the Amacrine cells (corresponding to the second low-pass filtering).

To efficiently simulate these low-level processes, we use a Daubechies-4 wavelet to decompose each of the six preattentive features into 4 scales. The process is illustrated in the lower part of Fig. 5.3. In each scale, the input feature map is gradually smoothed with a low-pass filter and down-sampled by a factor of 2. The smoothed version in the m-th scale is denoted as $A_{km}, m \in \{0, 1, 2, 3\}$. Meanwhile, the high-frequency components are extracted in three directions by high-pass filtering, denoted as H_{km}, V_{km} and D_{km}. Since the trivial details (usually noises) are beyond our concern, we only extract the local descriptors in the high-frequency sub-bands in the last three coarse layers. For the location (x, y) in the sub-band $S_{km} \in \{H_{km}, V_{km}, D_{km}\}_{m \in \{1,2,3\}}$, we compute the local energy as:

$$R_{km}(x, y) = S_{km}(x, y)^2 * G(0, \sigma), \tag{5.3}$$

where $G(0, \sigma)$ is a Gaussian kernel ($\sigma = 15$) to average the local energies. For convenience, we divide each scene into $N_r \times N_c$ macro-blocks (e.g., each block covers 16×16 pixels for a video with the resolution 640×480). Then all the energy maps are scaled to the size of $N_r \times N_c$, denoted as $R_{k1}, \ldots, R_{kL}, L = 6 \times 3 \times 3 = 54$. Thus the n-th macro-block in the scene can be characterized with a local descriptor $\mathbf{x}_{kn}$ with L components, where $n = 1, \ldots, N$ and $N = N_r \times N_c$. For the sake of simplicity, we normalize each dimension of the local descriptor into $[0, 1]$.

Beyond assigning a local descriptor to each block, we also extract a global descriptor to characterize the whole scene. Here we simply quantify the elements in each of the six preattentive features into $N_H = 8$ bins and get six feature histograms. These six histograms are combined to form a global descriptor H_k (with 48 components) to roughly represent the global characteristics of the k-th scene.

5.2.1.3   The Bottom-Up Saliency

After the filtering processes of the input scene, the irregular signals in various visual attributes can be extracted. These irregular signals are then transferred

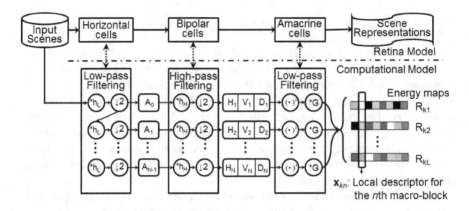

Fig. 5.3 The flowchart for extracting scene representations. Reproduced from (Li et al, 2010), with kind permission from Springer Science+Business Media.

to the higher levels of human vision system for estimating the bottom-up (and top-down) saliency. Typically, the bottom-up saliency can be computed through feature integration and unbiased competition. Since each dimension of a local descriptor is normalized to the same range, the integration can be carried out by summing up these dimensions with equal weights:

$$b_{kn} = \|\mathbf{x}_{kn}\|_1 / L. \tag{5.4}$$

With the bottom-up saliency b_{kn} for each block, we can obtain a bottom-up saliency map B_k. To efficiently localize the most salient locations in this map, the unbiased competition is then carried out by iteratively convolving the bottom-up saliency map B_k with a Difference of Gaussian filter. Here we simulate the competition process using the approach proposed in (Itti, 2000), that is:

$$B_k \leftarrow B_k + B_k * (c_{in} G(0, \sigma_{in}) - c_{out} G(0, \sigma_{out})), \tag{5.5}$$

where $G(0, \sigma_{in})$ and $G(0, \sigma_{out})$ are two Gaussian filters ($c_{in} = 0.25$ and $c_{out} = 0.8$ are two constants, $\sigma_{in} = 3$ and $\sigma_{out} = 7$). By iteratively carrying out the operation in (5.5), the distractors can be effectively suppressed while only the most salient locations are "pop-out". Since the integration process in (5.4) is carried out by integrating all features with equal weights, the competition process will not be biased to any specific visual feature. In this study, we empirically carry out the operation in (5.5) for three times. After that, the bottom-up saliency map for each scene is then normalized to the range of [0, 1], such that the most salient region in a scene reaches a saliency of 1.

5.2.2   The Probabilistic Multi-task Learning Framework

In this Section, we will introduce a probabilistic framework to incorporate the stimulus-driven and task-related components for saliency computation. Under this framework, a multi-task learning model and its algorithm will be presented to learn the task-related control for each scene.

5.2.2.1   Why Learnable?

Typically, the visual world is highly structured and scenes in videos are composed of features that are not random. For scenes with similar contents, neurobiological studies indicate that visual neurons exhibit tuning properties that can be optimized to respond to recurring features in the visual input (Chun, 2005). It is believed that the stable properties of the visual environment can be encoded and consulted to guide the visual processing. The stable properties, such as rough spatial layout and predicable variations, work as the contextual priors for individuals to find the target in similar environments. When encountering scenes with similar stable properties, people often avoid spending more neural resources in spatial locations with lower probabilities of containing the target. In neurobiology, this phenomenon is called contextual cueing effect (Itti et al, 2005).

From the computational perspective, we represent the stable properties of a scene with the global scene descriptor. More importantly, we assume that **scenes with similar global descriptors will undergo the similar processes in visual saliency computation**. As a consequence, the contextual cueing effect can be represented as adopting similar task-related "stimulus-saliency" functions and model fusion strategies in similar scenes. In this study, these functions and fusion strategies are learnt from scenes in the training data and then transferred to new scenes for estimating their visual saliency maps.

Typically, saliency can be obtained by combining the outputs of bottom-up model and various task-related models. However, since the output of each task-related model is unknown in the training data, it is infeasible to infer each task-related "stimulus-saliency" function independently. Usually, the tasks on different scenes are inherently correlated (e.g., tasks on consecutive frames). Therefore, **these task-related models, as well as their correlations, should be considered simultaneously in a multi-task learning framework**. By an appropriate sharing of information across tasks, each task may benefit from the others and their corresponding "stimulus-saliency" functions can be efficiently inferred from the same training data.

5.2.2.2   The Probabilistic Framework

Given a scene with rich contents, different people may focus on different locations. Their attention is driven either by the visual attributes of the stimuli or by the tasks. Let E_{kn} be the event that the n-th macro block is the salient block in the k-th scene. Thus we can formulate the generation of visual saliency in a probabilistic framework as:

$$p\left(E_{kn}|\mathbf{x}_{kn}\right) = \sum_{t=1}^{T_k} p\left(\mathcal{T}_{kt}\right) p\left(E_{kn}|\mathbf{x}_{kn}, \mathcal{T}_{kt}\right) + p\left(\mathcal{T}_{kb}\right) p\left(E_{kn}|\mathbf{x}_{kn}, \mathcal{T}_{kb}\right), \quad (5.6)$$

where $p\left(\mathcal{T}_{kt}\right)$ (or $p\left(\mathcal{T}_{kb}\right)$) denotes the probability that the task $\mathcal{T}_{kt}$ (or bottom-up process $\mathcal{T}_{kb}$) controls the deployment of attention in the k-th scene, and $p\left(E_{kn}|\mathbf{x}_{kn}, \mathcal{T}_{kt}\right)$ (or $p\left(E_{kn}|\mathbf{x}_{kn}, \mathcal{T}_{kb}\right)$) denotes the probability that the n-th block, which is represented by the local descriptor $\mathbf{x}_{kn}$, is the salient block when performing the task $\mathcal{T}_{kt}$ (or $\mathcal{T}_{kb}$). Note that here T_k is the number of tasks adopted on the k-th scene. Thus in this probabilistic framework, the problem of visual saliency computation is modeled by taking the stimulus-driven and task-related factors into account simultaneously.

However, it is difficult to estimate $p\left(E_{kn}|\mathbf{x}_{kn}, \mathcal{T}_{kt}\right)$ directly from the training data since the adopted tasks are unknown. To solve this problem, let $f_{kt}\left(\mathbf{x}\right)$ be the "stimulus-saliency" function relevant to the task $\mathcal{T}_{kt}$, and set $y_{kn}=p\left(E_{kn}|\mathbf{x}_{kn}\right)$ and $f_{kt}\left(\mathbf{x}_{kn}\right) = p\left(E_{kn}|\mathbf{x}_{kn}, \mathcal{T}_{kt}\right)$. Then we can re-write (5.6) by setting $\alpha_{kt} = p\left(\mathcal{T}_{kt}\right)$, $\alpha_{kb} = p\left(\mathcal{T}_{kb}\right)$ and $b_{kn} = p\left(E_{kn}|\mathbf{x}_{kn}, \mathcal{T}_{kb}\right)$:

$$y_{kn} \leftarrow e_{kn} = \sum_{t=1}^{T_k} \alpha_{kt} f_{kt}\left(\mathbf{x}_{kn}\right) + \alpha_{kb} b_{kn}, \qquad (5.7)$$

where $0 \leq \alpha_{kt}, \alpha_{kb} \leq 1$ and $\sum_{t=1}^{T_k} \alpha_{kt} + \alpha_{kb} = 1$. Note that here e_{kn} is the predicted saliency which is expected to approximate y_{kn}. Given the eye-fixation or labeled salient region data, we can record the attended locations of all the viewers and then get a "ground-truth" saliency map for each scene. Thus in the training stage, y_{kn} can be obtained by calculating the distribution of the attended locations. For example, y_{kn} can be calculated as the number of eye-fixations received by the n-th block, which is then normalized to [0, 1]. Given the parameters $\{\alpha_{kt}, \alpha_{kb}\}$ and the "stimulus-saliency" functions $\{f_{kt}\left(\mathbf{x}\right)\}$ that are learned from the training data, the saliency computation can then be transformed into a multivariate regression problem with the input $\mathbf{x}_{kn}$. In the following two subsections, we will discuss in detail how to model and learn these parameters and the "stimulus-saliency" functions in a multi-task learning framework.

In this study, we will learn a set of "stimulus-saliency" functions and a fusion strategy for each scene. Thus another problem is how to determine the tasks and fusion strategy for a new scene. Recall that scenes with similar global descriptors are supposed to undergo similar visual processes. Then we can train a scene classifier to select "stimulus-saliency" functions and fusion strategies for the new scenes based on the global descriptors. In this study, we define the similarity of the u-th scene and the v-th scene with respect to their global descriptors H_u and H_v:

$$S_{uv} = \frac{H_u^{\mathrm{T}} H_v}{\|H_u\|_2 \|H_v\|_2}, \tag{5.8}$$

Based on this similarity, we adopt a KNN classifier (1-NN in this study) to find the most similar scenes in the training data, and use the same "stimulus-saliency" functions and fusion strategy to estimate the visual saliency on the new scene. After the block-level visual saliency computation, we can then obtain a saliency map that represents the distribution of salient objects. To "pop-out" the most salient location, we revise the predicted saliency by:

$$\widehat{e}_{kn} = G(0, \sigma) * e_{kn}^{\beta}, \tag{5.9}$$

where the exponential operation e_{kn}^{β} with $\beta > 1$ is used to suppress the saliency in distractors while maintaining the most salient region ($\beta = 2$ in this study). Here a Gaussian smooth ($\sigma = 5$) is used to ensure that high saliency values are also assigned to the regions around the most salient location (e.g., the internal region of the target). Likewise, the revised saliency map is then normalized to $[0, 1]$.

5.2.2.3   Multi-task Learning Model

Given K training scenes each with N blocks, we have $K \times N$ input-output pairs $(\mathbf{x}_{kn}, y_{kn})$ with $k \in \{1, \dots, K\}$ and $n \in \{1, \dots, N\}$. Thus the "stimulus-saliency" functions as well as the combination weights can be inferred by solving the problem:

$$\min_{\{f_{kt}, \alpha_{kt}, \alpha_{kb}\}} \frac{1}{KN} \sum_{k=1}^{K} \sum_{n=1}^{N} l \left(\sum_{t=1}^{T_k} \alpha_{kt} f_{kt}(\mathbf{x}_{kn}) + \alpha_{kb} b_{kn}, y_{kn} \right),$$

$$s.t. 0 \leq \alpha_{kt}, \alpha_{kb} \leq 1, \tag{5.10}$$

$$\sum_{t=1}^{T_k} \alpha_{kt} + \alpha_{kb} = 1, \forall k \in \{1, \dots, K\},$$

where $l(\cdot)$ is a predefined loss function which quantifies the cost of predicting saliency for the input $\mathbf{x}_{kn}$ with the "ground-truth" y_{kn}.

Unfortunately, (5.10) is usually intractable since there are too many parameters to optimize. To solve this problem, we can decompose a complex task (i.e., a complex visual search task in this study) into the conjunction of several simpler tasks (Wolfe, 1998). For example, as shown in Fig. 5.4, searching the black vertical bar can be viewed as a conjunction of searching the black bar and searching the vertical bar. Thus we assume that there are a set of typical tasks, denoted as $\left\{\hat{\mathcal{T}}_t\right\}$, and every complex task can be approximated by the conjunction of these typical tasks.

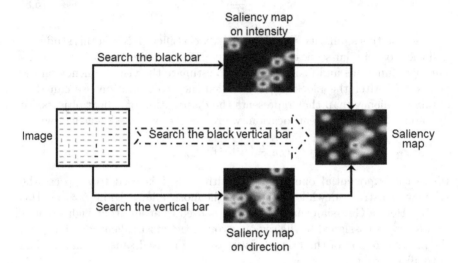

Fig. 5.4 A complex search task can be decomposed to several simpler tasks. Reproduced from (Li et al, 2010), with kind permission from Springer Science+Business Media.

With this assumption, we have:

$$p\left(E_{kn}|\mathbf{x}_{kn}, \mathcal{T}_{ki}\right) = \sum_{t \in \mathbb{T}} p\left(\hat{\mathcal{T}}_t|\mathcal{T}_{ki}\right) p\left(E_{kn}|\mathbf{x}_{kn}, \hat{\mathcal{T}}_t\right), \qquad (5.11)$$

where $\mathbb{T}$ denotes the set of indices of typical tasks.

For the sake of simplicity, we let $|\mathbb{T}|$ be the number of typical tasks and set $\beta_{kit} = p\left(\hat{\mathcal{T}}_t|\mathcal{T}_{ki}\right)$ the probability that the complex task $\mathcal{T}_{ki}$ comprises of the typical task $\hat{\mathcal{T}}_t$, and $\hat{f}_t\left(\mathbf{x}_{kn}\right) = p\left(E_{kn}|\mathbf{x}_{kn}, \hat{\mathcal{T}}_t\right)$ be a typical "stimulus-saliency" function. We can then rewrite (5.11) by approximating $f_{ki}\left(\mathbf{x}\right)$ with a linear combination of the typical "stimulus-saliency" functions:

$$f_{ki}\left(\mathbf{x}\right) = \sum_{t \in \mathbb{T}} \beta_{kit} \hat{f}_t\left(\mathbf{x}\right), \qquad (5.12)$$

where $\sum_{t \in \mathbb{T}} \beta_{kit} = 1$ and $0 \le \beta_{kit} \le 1$. Thus by incorporating (5.12) into (5.10), the optimization problem can be re-formulated as:

$$\min_{\mathbb{F},\mathbb{W}} \frac{1}{KN} \sum_{k=1}^{K} \sum_{n=1}^{N} l \left(\sum_{t \in \mathbb{T}} c_{kt} \hat{f}_t \left(\mathbf{x}_{kn} \right) + \alpha_{kb} b_{kn}, y_{kn} \right),$$

$$s.t. 0 \le c_{kt}, \alpha_{kb} \le 1, \tag{5.13}$$

$$\sum_{t \in \mathbb{T}} c_{kt} + \alpha_{kb} = 1, \forall k \in \{1, \ldots, K\},$$

where $c_{kt} = \sum_{i=1}^{T_k} \alpha_{ki} \beta_{kit}$. Here we let $\mathbf{c}_k = \left(c_{k1}, \ldots, c_{k|\mathbb{T}|}, \alpha_{kb} \right)^{\mathbf{T}}$ be a column vector with $|\mathbb{T}| + 1$ components. The sets of typical "stimulus-saliency" functions and combination weights are denoted as $\mathbb{F} = \left\{ \hat{f}_t(\mathbf{x}) \right\}_{t \in \mathbb{T}}$ and $\mathbb{W} = \{\mathbf{c}_k\}_{k \in \{1, \ldots, K\}}$ respectively. For convenience, we set the average prediction error for all the training samples in the K scenes as:

$$\mathcal{L}\left(\mathbb{F}, \mathbb{W}\right) = \frac{1}{KN} \sum_{k=1}^{K} \sum_{n=1}^{N} l \left(\sum_{t \in \mathbb{T}} c_{kt} \hat{f}_t \left(\mathbf{x}_{kn} \right) + \alpha_{kb} b_{kn}, y_{kn} \right). \tag{5.14}$$

Similar to other multi-task learning algorithms (Evgeniou et al, 2005; Argyriou et al, 2007; Jacob et al, 2009), we introduce a penalty term $\Omega\left(\mathbb{F}, \mathbb{W}\right)$ which can be used to improve the generalization ability of the learnt "stimulus-saliency" functions and fusion strategies. That is, the typical tasks are expected to diversify from each other. In general, two typical tasks are expected to have different targets, thus focusing on different visual attributes. Different task-related "stimulus-saliency" functions are then expected to give diversified outputs on every input sample. For simplification, we define the inner product of two functions as $\left\langle \hat{f}_u, \hat{f}_v \right\rangle = \int_{\mathbf{x}} \hat{f}_u(\mathbf{x}) \hat{f}_v(\mathbf{x}) d\mathbf{x}$ and the difference of two functions can be defined as:

$$D\left(\hat{f}_u, \hat{f}_v \right) = \left\langle \hat{f}_u - \hat{f}_v, \hat{f}_u - \hat{f}_v \right\rangle. \tag{5.15}$$

Thus we can define the penalty term on the energy of difference between two "stimulus-saliency" functions:

$$\Omega_{inter}\left(\mathbb{F}, \mathbb{W}\right) = -\frac{1}{|\mathbb{T}|^2 - |\mathbb{T}|} \sum_{u \ne v}^{|\mathbb{T}|} D\left(\hat{f}_u, \hat{f}_v \right). \tag{5.16}$$

However, we can see that the penalty term $\Omega_{inter}\left(\mathbb{F}, \mathbb{W}\right)$ has no lower bound since the output ranges of the "stimulus-saliency" functions are not bounded. Therefore, it is reasonable to set a penalty term on the outputs of the "stimulus-saliency" functions:

$$\Omega_{rov}\left(\mathbb{F}, \mathbb{W}\right) = \frac{1}{|\mathbb{T}|} \sum_{t \in \mathbb{T}} \left\langle \hat{f}_t, \hat{f}_t \right\rangle. \tag{5.17}$$

Moreover, video is not a randomized copy of natural scenes. Successive scenes in a shot usually have similar targets and distractors. For these scenes, the fusion strategies can be strongly correlated. Based on the similarity definition in (5.8), we set a penalty term on the fusion strategies:

$$\Omega_{cont}\left(\mathbb{F}, \mathbb{W}\right) = \frac{1}{K^2 - K} \sum_{u \neq v}^{K} S_{uv} \cdot \|\mathbf{c}_u - \mathbf{c}_v\|_2^2 \cdot e^{-\frac{(u-v)^2}{\sigma}}, \qquad (5.18)$$

where the $\exp\left(\cdot\right)$ term is used as a temporal constraint to avoid computing the correlations between two frames that are far away ($\sigma = 5$). We can see that the fusion strategy on a video frame is more susceptible to the fusion strategies adopted in adjacent similar video frames.

For convenience, we denote the three penalties as Ω_{inter}, Ω_{rov} and Ω_{cont}, respectively. In the optimization, the overall penalty can be rewritten as:

$$\Omega\left(\mathbb{F}, \mathbb{W}\right) = \epsilon_{inter}\Omega_{inter} + \epsilon_{rov}\Omega_{rov} + \epsilon_{cont}\Omega_{cont}, \qquad (5.19)$$

where $\epsilon_{inter}, \epsilon_{rov}$ and ϵ_{cont} are three non-negative weights that balance the influences of different penalty terms (in this study, these three weights are selected using the validation set). Moreover, we should select these weights to ensure that the overall penalty has a lower bound (e.g., increasing ϵ_{rov} or decreasing ϵ_{inter}). By incorporating the average prediction error (5.14) and the penalty term (6.7) into a uniform optimization objective, the problem is equivalent to a non-convex minimization formulation:

$$\min_{\mathbb{F}, \mathbb{W}} \mathcal{L}\left(\mathbb{F}, \mathbb{W}\right) + \lambda\Omega\left(\mathbb{F}, \mathbb{W}\right),$$
$$s.t. 0 \preceq \mathbf{c}_k \preceq 1, \qquad (5.20)$$
$$\|\mathbf{c}_k\|_1 = 1, \forall k \in \{1, \ldots, K\}.$$

5.2.2.4   The Learning Algorithm

Note that when optimizing (5.20), the combination weights $\mathbb{W}$ are dependent on the "stimulus-saliency" functions $\mathbb{F}$ and vice versa, which prevents a straightforward solution. To solve this problem, we employ EM-like iterations, i.e., alternating between estimating the combination weights $\mathbb{W}$ and optimizing the "stimulus-saliency" functions $\mathbb{F}$. Without loss of generality, we use the square of prediction errors as the loss function $l(\cdot)$ and approximate the inner product of two functions by:

$$\left\langle \hat{f}_u, \hat{f}_v \right\rangle = \int_{\mathbf{x}} \hat{f}_u(\mathbf{x})\hat{f}_v(\mathbf{x})d\mathbf{x} \approx \frac{1}{KN} \sum_{k=1}^{K} \sum_{n=1}^{N} \hat{f}_u(\mathbf{x}_{kn})\hat{f}_v(\mathbf{x}_{kn}). \qquad (5.21)$$

Here we suppose that the "stimulus-saliency" functions in $\mathbb{F}$ are one-order differentiable and are controlled by a set of hyper-parameters $\Pi = \{\pi_t\}_{t\in\mathbb{T}}$. Thus we can initialize $\mathbb{W}$ and Π with random values, and then iteratively update $\mathbb{W}$ and Π. For the sake of simplicity, we denote $x^{(\triangle)} = x^{(m)}$ if parameter x has been updated in the m-th iteration, otherwise $x^{(\triangle)} = x^{(m-1)}$. Thus the optimization in the m-th iteration can be performed as:

- E-Step: update $\mathbf{c}_k$ by solving the problem which contains only the terms in (5.20) that are related to $\mathbf{c}_k$:

$$
\begin{aligned}
\mathbf{c}_k^{(m)*} = \arg\min_{\mathbf{c}_k} \frac{1}{KN} \sum_{n=1}^{N} \left\| \mathbf{v}_{kn}^{\mathbf{T}} \mathbf{c}_k - y_{kn} \right\|_2^2 \\
+ \frac{\lambda \epsilon_{cont}}{K^2 - K} \sum_{u\neq k}^{K} S_{uk} \cdot \left\| \mathbf{c}_u^{(\triangle)} - \mathbf{c}_k \right\|_2^2 \cdot e^{-\frac{(u-k)^2}{\sigma}} \\
s.t. 0 \preceq \mathbf{c}_k \preceq 1, \\
\left\| \mathbf{c}_k \right\|_1 = 1,
\end{aligned}
\tag{5.22}
$$

where $\mathbf{v}_{kn} = \left(\hat{f}_1^{(m-1)}(\mathbf{x}_{kn}), \dots, \hat{f}_{|\mathbb{T}|}^{(m-1)}(\mathbf{x}_{kn}), b_{kn} \right)^{\mathbf{T}}$ is a column vector with $|\mathbb{T}| + 1$ components which consists of the predicted saliency values of bottom-up and task-relevant models in previous iteration. We can see that this problem can be efficiently solved using quadratic programming.

- M-Step: in the m-th iteration, update π_t by solving the problem which contains only the terms in (5.20) that are related to π_t:

$$
\begin{aligned}
\pi_t^{(m)*} = \arg\min_{\pi_t} \frac{1}{KN} \sum_{k=1}^{K} \sum_{n=1}^{N} \Big\{ \left[c_{kt}^{(m)} \hat{f}_t(\mathbf{x}_{kn}) + \eta_{kn} \right]^2 \\
- \frac{\lambda \epsilon_{inter}}{|\mathbb{T}|^2 - |\mathbb{T}|} \sum_{u\neq t}^{|\mathbb{T}|} \left[\hat{f}_u^{(\triangle)}(\mathbf{x}_{kn}) - \hat{f}_t(\mathbf{x}_{kn}) \right]^2 \\
+ \frac{\lambda \epsilon_{rov}}{|\mathbb{T}|} \left[\hat{f}_t(\mathbf{x}_{kn}) \right]^2 \Big\},
\end{aligned}
\tag{5.23}
$$

where η_{kn} is calculated as:

$$
\eta_{kn} = \sum_{u\neq t}^{|\mathbb{T}|} c_{ku}^{(m)} \hat{f}_u^{(\triangle)}(\mathbf{x}_{kn}) + \alpha_{kb}^{(m)} b_{kn} - y_{kn}.
\tag{5.24}
$$

In solving (5.23), the typical "stimulus-saliency" function $\hat{f}_t(\mathbf{x}_{kn})$ can use any one-order differentiable linear or nonlinear functions. Recall that we have $\hat{f}_t(\mathbf{x}_{kn}) = p\left(E_{kn}|\mathbf{x}_{kn}, \hat{\mathcal{J}}_t \right)$. Therefore, the output of the function should be in the range $[0, 1]$. For example, the linear "stimulus-saliency" function can adopt the following form:

$$\hat{f}_t(\mathbf{x}) = \omega_t^{\mathbf{T}} \mathbf{x}, 0 \preceq \omega_t \preceq 1 \text{ and } \|\omega_t\|_1 = 1. \tag{5.25}$$

When using this linear function, the problem in (5.23) can be efficiently solved by quadratic programming. We can also use the radial basis function (RBF) as the nonlinear "stimulus-saliency" function:

$$\hat{f}_t(\mathbf{x}) = e^{-\gamma_t \|\mathbf{x} - \omega_t\|_2^2}, \gamma_t > 0. \tag{5.26}$$

In this case, the optimization problem in (5.23) can be solved using the gradient-descent based method. Note that here the "stimulus-saliency" function corresponding to each complex task in (5.12) is now approximated by a RBF network (i.e., a simple neural network), making the task-related model **biologically plausible**. Similarly, we can adopt any other nonlinear function (e.g., the sigmoid function) and solve the optimization problem using the gradient-descent based method.

By iteratively updating the combination weights $\mathbb{W}$ and the "stimulus-saliency" functions $\mathbb{F}$, we have a decreasing overall prediction error:

$$\begin{aligned}
&\mathcal{L}\left(\mathbb{F}^{(m)}, \mathbb{W}^{(m)}\right) + \lambda \Omega \left(\mathbb{F}^{(m)}, \mathbb{W}^{(m)}\right) \\
&\leq \mathcal{L}\left(\mathbb{F}^{(m-1)}, \mathbb{W}^{(m-1)}\right) + \lambda \Omega \left(\mathbb{F}^{(m-1)}, \mathbb{W}^{(m-1)}\right).
\end{aligned} \tag{5.27}$$

By selecting proper ϵ_{rov} and ϵ_{inter}, the optimization objective has a lower-bound. Therefore, the EM-like algorithm will iterate until the loss function in (5.20) converges, or a certain number of iterations is reached. To illustrate the effectiveness of the algorithm, we use the linear and nonlinear "stimulus-saliency" function in (5.25) and (5.26) in the experiments. Experimental results verify that the algorithm will converge in about 50 iterations.

5.2.3   Experiments

In this Section we evaluate the proposed approach on three public datasets to validate whether the learned saliency model could capture the salient locations in video sequences. The three datasets are denoted as **ORIG**, **MTV** and **RSD**, where the **ORIG** and **MTV** data-sets are two eye-fixation datasets proposed in (Itti, 2008), and the **RSD** dataset (Li et al, 2009) is a video dataset with manually labeled regional saliency. Details of the three datasets are summarized as follows:

1. **ORIG**: The **ORIG** dataset consists of over 46,000 video frames in 50 video clips (25 minutes). The dataset contains scenes such as "outdoors day &

night," "crowds," "TV news," "sports," "commercials," "video games" and "test stimuli." The eye traces of 8 subjects when watching these clips (4 to 6 subjects/clip) were recorded using a 240HZ ISCAN RK-464 eye-tracker. Since these video clips cover different genres, we use this dataset to evaluate whether the models trained from some genres (e.g., video games) can be generalized to other genres (e.g., sports and TV news).

2. **MTV**: The **MTV** dataset contains 50 video clips, and is constructed by cutting the video clips in the **ORIG** dataset into 1-3s "clippets" and reassembling those clippets in random order. The eye-traces were recorded in a similar way as the **ORIG** dataset, but with 8 different subjects. Since these video clips have similar contents, we use this dataset to evaluate whether the model trained on some scenes can remain effective in similar scenes.

3. **RSD**: The **RSD** dataset contains 431 videos with a total length of 7.5 hours. In total, 764,806 frames are involved. The dataset mainly covers videos from six genres: documentary, ad, cartoon, news, movie and surveillance. In these videos, 62,356 key frames are selected. Then 23 subjects are assigned to manually label the salient regions with one or multiple rectangles. On this dataset, we aim at evaluating whether the proposed approach can be used to effectively estimate regional saliency.

Based on the recorded eye traces or labeled regions, we generate a ground-truth saliency map (GSM) for each video frame by combining the eye-fixations or labeled rectangles of all subjects. Similar to the algorithms (Hou and Zhang, 2007; Guo et al, 2008), the GSMs are further convolved with a 2D Gaussian kernel ($\sigma = 3$) to smooth the edges. On these datasets, eight state-of-the-art algorithms are used for comparison:

- Itti98 (Itti et al, 1998) and Itti01 (Itti and Koch, 2001a): Two approaches that adopt the typical center-surround framework;
- Itti05 (Itti and Baldi, 2005): A novelty-based approach;
- Harel07 (Harel et al, 2007): A graph-based approach;
- Hou07 (Hou and Zhang, 2007) and Guo08 (Guo et al, 2008): Two approaches based on spectrum analysis;
- Zhai06 (Zhai and Shah, 2006): An approach based on irregular motion and
- Peters07 (Peters and Itti, 2007): A learning-based approach that learns a unified "stimulus-saliency" projection matrix for all scenes.

In the experiment, these algorithms will be compared with the wavelet-based bottom-up model (PMTL-WBU) and the probabilistic multi-task learning model with linear function (PMTL-Linear) or nonlinear function (PMTL-RBF). In the comparison, we adopt the receiver operator characteristic (**ROC**) to evaluate the similarity between the estimated saliency maps (ESMs) and ground-truth saliency maps (GSMs). Different thresholds are adopted to select salient locations from ESMs, which are then validated by GSMs. After that, the **ROC** curve is plotted as the *false positive rate* vs. *true positive rate*, and the **AUC** score is defined as the area under the **ROC** curve. Perfect prediction corresponds to a score of 1 and random prediction corresponds to a score of 0.5. Moreover, we also compute the improvement (on **AUC**) of PMTL-Linear against all the other approaches and illustrate the **ROC** curves to demonstrate the performance of these approaches in detail.

5.2.3.1   Results

This Section presents the performance of different algorithms on the three datasets. First, we carried out the effectiveness test on the **MTV** dataset to evaluate whether the leant "stimulus-saliency" functions and fusion strategies can be effectively transferred to similar scenes. The second experiment is carried out on the **ORIG** dataset, which is designed to evaluate whether the learnt "stimulus-saliency" functions and fusion strategies can be generalized to different video genres. Finally, the third experiment is performed on the **RSD** dataset to evaluate whether the proposed algorithm can work well on the regional saliency dataset.

5.2.3.2   Effectiveness Test on the MTV Dataset

Here we use the **MTV** dataset for effectiveness evaluation since the video clips have similar contents. We randomly partitioned the dataset into 4 subsets equally. At each time, we select one subset as the training set for PMTL-Linear, PMTL-RBF and Peters07, one subset as the validation set for optimal parameter selection (e.g., the number of typical tasks and the weights of various penalty terms for PMTL-Linear and PMTL-RBF) and the remaining two subsets as the testing set. The results are shown in Table 5.1. Some representative results are given in Fig. 5.5, and the **ROC** curves are illustrated in Fig. 5.6.

Table 5.1 Performance of various approaches on **MTV**. Reproduced from (Li et al, 2010), with kind permission from Springer Science+Business Media.

Algorithm		**AUC**	Improvement (%)
	Itti98	0.54	39.60
	Itti01	0.53	40.20
	Itti05	0.59	27.72
Bottom-Up	Hou07	0.62	20.38
	Guo08	0.63	19.28
	Harel07	0.54	37.62
	Zhai06	0.62	20.11
	PMTL-WBU	0.71	6.23
	Peters07	0.65	14.54
Top-Down	PMTL-RBF	0.72	3.99
	PMTL-Linear	**0.75**	

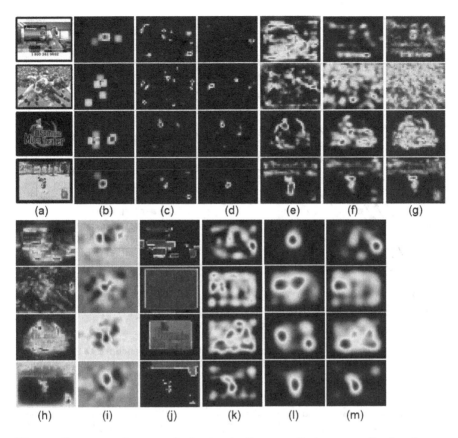

(a)   (b)   (c)   (d)   (e)   (f)   (g)

(h)   (i)   (j)   (k)   (l)   (m)

Fig. 5.5 Representative examples in visual saliency prediction on eye-fixation dataset **MTV**. (a) Original frames; (b) fixation density maps; (c) Itti98; (d) Itti01; (e) Itti05; (f) Hou07; (g) Guo08; (h) Harel07; (i) Peters07; (j) Zhai06; (k) PMTL-WBU; (l) PMTL-Linear; (m) PMTL-RBF. Reproduced from (Li et al, 2010), with kind permission from Springer Science+Business Media.

Among all bottom-up models, PMTL-WBU outperforms over the other approaches. We can see that the Itti98 and Itti01 yield the lowest **AUC** scores since they only take care of the most salient regions using the "winner-take-all" competition (as shown in Fig. 5.5(c) and Fig. 5.5(d)). Harel07 obtains comparative performance since only the less-visited pixels are selected in the random walk process. This algorithm often fails to detect the internal part of the target (as shown in Fig. 5.5(h)). By focusing on the inter-frame variations, Zhai06 and Itti05 have achieved a bit improvement. However, Zhai06 can hardly detect small salient objects or objects with non-rigid motion (as shown in Fig. 5.5(j)), while Itti05 fails in static scenes or scenes with camera motion (as shown in Fig. 5.5(e)). Alternatively, Hou07, Guo08 and PMTL-WBU can well detect the irregularities in the transform domain. In these methods, the salient locations will rarely be missed, leading to high **AUC**

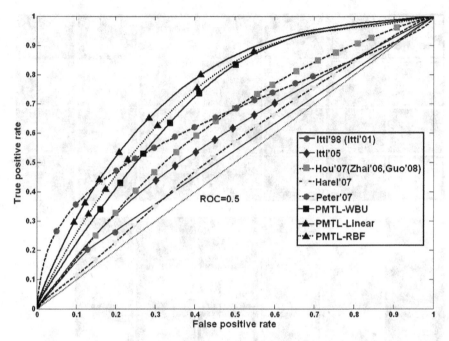

Fig. 5.6 ROC curves of different algorithms on the **MTV** benchmark. Reproduced from (Li et al, 2010), with kind permission from Springer Science+Business Media.

scores. However, Hou07 and Guo08 may have difficulties to suppress the distractors (as shown in Fig. 5.5(f) and Fig. 5.5(g)). In contrast, PMTL-WBU detects the irregularities from multiple scales and directions. In this process, the targets will be enhanced multiple times, leading to a higher **AUC**=0.71.

On average, the three learning-based approaches, including PMTL-Linear, PMTL-RBF and Peters07, have achieved impressive performance compared with the bottom-up approaches. As shown in Fig. 5.5(i), Peters07 could effectively select the most salient regions. However, this method adopts a unified model for all scenes and thus tends to give "mean" predictions. In contrast, PMTL-Linear and PMTL-RBF obtain the best performance (as shown in Fig. 5.6). This is mainly due to two reasons. First, PMTL-WBU is adaptively incorporated into PMTL-Linear and PMTL-RBF to ensure that no salient objects are missed. Second, the top-down components elegantly select the probable tasks for different scenes, thus using proper "stimulus-saliency" functions and model fusion strategies to emphasize **ONLY** the salient targets. Therefore, it can be safely concluded that the learning algorithm is effective to transfer the experience from past scenes to similar new scenes for visual saliency computation.

Surprisingly, we can see from Table 5.1 that PMTL-RBF achieves a lower **AUC** than PMTL-Linear. The reason is that the radial basis function performs slightly weaker than the linear function in distinguishing targets from distractors. For example, a local descriptor with all the components equal to zero (e.g., the local descriptor corresponding to the smooth background region) will always guarantee a zero output in the linear function. However, this fact may not hold for the radial basis function. From Fig. 5.5(l) and Fig. 5.5(m), we can see that PMTL-RBF is less effective in suppressing the distractors than PMTL-Linear, leading to a lower **ROC** curve (as shown in Fig. 5.6). A probable solution to this problem is to further represent each typical "stimulus-saliency" function with a RBF network (i.e., a multi-layer neural network for each complex visual search task). In this case, the distractors might be suppressed more effectively.

5.2.3.3   Robustness Test on the ORIG Dataset

In this experiment, we evaluate the robustness of the proposed approach. That is, we want to explore whether the "stimulus-saliency" functions and fusion strategies trained on some genres (e.g., TV games, news) can be applied to different genres (e.g., sports, advertisements). Similar to the previous experiment, we randomly partitioned the **ORIG** dataset into 4 subsets equally, in which each subset covers different video genres. At each time, we select two subsets as the training set and the validation set for PMTL-Linear, PMTL-RBF and Peters07, respectively. The remaining two subsets are then used as the testing set. The results are listed in Table 5.2[1], and their **ROC** curves are illustrated in Fig. 5.7.

Table 5.2 Performance of various approaches on **ORIG**. Reproduced from (Li et al, 2010), with kind permission from Springer Science+Business Media.

Algorithm		**AUC**	Improvement (%)
	Itti98	0.56	35.96
	Itti01	0.55	36.82
	Itti05	0.62	22.01
Bottom-Up	Hou07	0.66	13.86
	Guo08	0.67	12.61
	Harel07	0.58	29.80
	Zhai06	0.64	18.92
	PMTL-WBU	0.72	5.74
	Peters07	0.62	22.57
Top-Down	PMTL-RBF	0.74	2.30
	PMTL-Linear	**0.76**	

From Table 5.1 and 5.2, we can see that the performances of all the bottom-up approaches on the **ORIG** dataset are better than those on the **MTV**

---

[1] The **AUC** scores reported in Tables 5.2, 6.1 and 7.1 may change slightly due to different experimental settings (e.g., sizes of Gaussian kernels used in convolution, resolutions of saliency maps and settings of training/validation/testing sets).

dataset. Surprisingly, the performance of the pure learning-based approach (Peters07) becomes worse. The possible reason is that in this experiment, the testing video clips cover different genres with the training clips; while in the previous experiment, the training and testing videos contain similar contents. Particularly, PMTL-Linear and PMTL-RBF demonstrates high robustness across the two datasets. The main reason is that the proposed approach contains both the bottom-up and the top-down components, in which the top-down component often plays a greater role than the bottom-up component when processing the conversant scenes and vice versa. In this experiment, PMTL-Linear achieves 22.6% improvement over Peters07 in **AUC**.

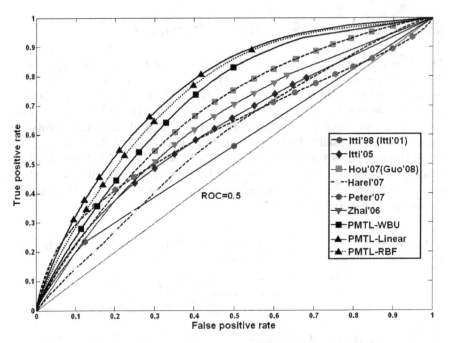

Fig. 5.7 ROC curves of different algorithms on the **ORIG** benchmark. Reproduced from (Li et al, 2010), with kind permission from Springer Science+Business Media.

Here we give a temporal analysis to explain why these bottom-up approaches perform better on the **ORIG** dataset than on the **MTV** dataset. As in (Harel et al, 2007; Marat et al, 2009), we first quantify the difference between viewers' attention by computing the mean inter-subject **ROC** on the **MTV** dataset. Let N_k be the number of viewers who have watched the k-th scene. We randomly select $N_k - 1$ viewers to derive an ESM from their fixations. The fixations of the left viewer are used to generate a GSM. The **ROC** of the ESM and GSM are then computed, denoted as "inter-subject **ROC**." This operation is repeated N_k times in each scene. Typically, a low inter-subject **ROC** indicates that there is a high divergence between different

viewers' attention. Moreover, we have manually labeled the shot cuts in the
MTV dataset. In total we get 639 shots with an average length of 51 ± 23
frames. Then we plot the mean inter-subject **ROC** value as a function of
the frame index after each shot cut. As shown in Fig. 5.8, the inter-subject
ROC is low after each shot cut. As proposed in (Marat et al, 2009), the
reason is that the fixations after each shot cut still stay at the previous positions for a few frames. After a short duration from the shot cut, viewer's
attention starts to be driven by the bottom-up process since the bottom-up
influence acts faster than the top-down one (Wolfe et al, 2000; Henderson,
2003). Since the bottom-up processes are stimulus-driven, viewers will start
to focus on the same salient location, leading to higher inter-subject **ROC**.
After 25-30 frames, the inter-subject **ROC** reaches its maximum. After that,
the top-down factors start to work to diversify viewers' attention, leading to
a decreasing inter-subject **ROC**. To remove the influence of shot cuts, we
calculate the **AUC** scores after the inter-subject **ROC** reaches its maximum
(about 25 frames from the beginning of a shot). In this case, the **AUC** scores
of the bottom-up approaches on the **MTV** dataset are almost the same with
those in the **ORIG** dataset.

Fig. 5.8 The inter-subject **ROC** on the **MTV** dataset. The horizontal axis is the number of frames from the shot cuts, while the vertical axis is inter-subject **ROC**. Reproduced from (Li et al, 2010), with kind permission from Springer Science+Business Media.

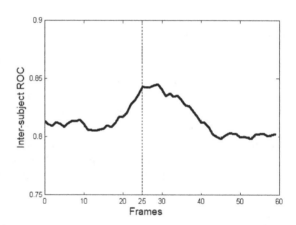

In addition, we also conduct an experiment by training and testing the
proposed approach on the same video genres. The objective is to evaluate
the usefulness of the top-down component. From the **ORIG** dataset, we
select six genres, including "outdoor," "game," "crowd," "news," "sport"
and "ad." On each of the six genre, we randomly divided the videos into
training, validation and testing sets, respectively. The results for each genre
are listed in Table 5.3. Without loss of generality, we only demonstrate the
performance of PMTL-Linear while PMTL-RBF presents similar results.

From Table 5.2 and Table 5.3, we can see that the performance of PMTL-
Linear on the same genres (except "news") are much better than that on
different genres. The main reason is that videos in the same genre usually

Table 5.3 Performance of PMTL-Linear on different video genres of **ORIG**. Reproduced from (Li et al, 2010), with kind permission from Springer Science+Business Media.

Genre	Outdoor	Game	Crowd	Sport	Ad	News
AUC	0.83	0.82	0.78	0.80	**0.84**	0.72

have similar targets and viewers can localize them through similar visual search processes. Therefore, the experience can be more useful to guide the search processes for the scenes in the same genre and the KNN classifier can be more accurate to select the similar scenes. Moreover, we can see that PMTL-Linear performs the best on the "ad" genre while the performance is the worst on the "news" genre. The reason is that the attractive targets in commercial videos are usually emphasized by the video provider. In contrast, the news videos often contain rich contents while different content may be attractive to different category of viewers. Therefore, it is much easier to distinguish the targets from distractors in commercial videos than in news videos.

5.2.3.4   Performance on the RSD Dataset

Besides the effectiveness and robustness tests, the third experiment is to evaluate whether PMTL-Linear and PMTL-RBF work well on the regional saliency dataset (i.e., whether they can detect the whole salient object when trained with *regional saliency*). We also randomly partitioned the **RSD** dataset into 4 subsets equally and follow the same way to construct the training/validation/testing sets as in the previous experiments. The results are given in Table 5.4 and some representative examples are also illustrated in Fig. 5.9. The corresponding **ROC** curves are also shown in Fig. 5.10.

Table 5.4 Performance of various approaches on **RSD**. Reproduced from (Li et al, 2010), with kind permission from Springer Science+Business Media.

Algorithm		**AUC**	Improvement (%)
	Itti98	0.54	58.75
	Itti01	0.55	54.31
	Itti05	0.62	37.85
Bottom-Up	Hou07	0.68	25.95
	Guo08	0.63	36.00
	Harel07	0.58	46.46
	Zhai06	0.63	35.30
	PMTL-WBU	0.73	35.30
	Peters07	**0.85**	0.0
Top-Down	PMTL-RBF	0.81	5.37
	PMTL-Linear	**0.85**	

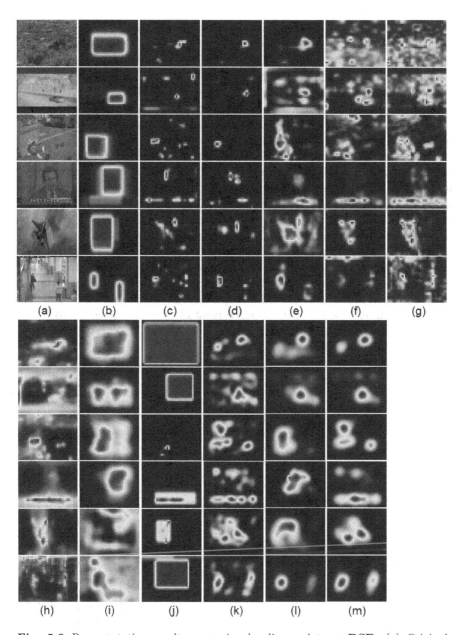

Fig. 5.9 Representative results on regional saliency dataset **RSD**. (a) Original frames; (b) labeled saliency maps; (c) Itti98; (d) Itti01; (e) Itti05; (f) Hou07; (g) Guo08; (h) Harel07; (i) Peters07; (j) Zhai06; (k) PMTL-WBU; (l) PMTL-Linear; (m) PMTL-RBF. Reproduced from (Li et al, 2010), with kind permission from Springer Science+Business Media.

From Table 5.4, we can see that the **AUC** scores for the bottom-up approaches are comparable with those in the **ORIG** and **MTV** datasets. In contrast, the learning-based methods have the highest **AUC** scores (0.85 for Peters07 and PMTL-Linear). Particularly, the **AUC** scores of PMTL-Linear and PMTL-RBF are remarkably higher than PMTL-WBU. As shown in Fig. 5.9(l), we can see that PMTL-Linear can effectively filter out the non-attractive regions which are also unique or rare in the scene (e.g., the apples in the third row, and the captions in the fourth row).

Moreover, we also notice that PMTL-Linear, PMTL-RBF and Peters07 have comparable **AUC** scores on the **RSD** dataset that are significantly higher than that on the **ORIG** and **MTV** datasets. The reason is that the learning problems are different in the eye-fixation dataset and the regional saliency dataset. Actually, in the **RSD** dataset the unselected blocks can be viewed as non-salient blocks and act as negative samples in the learning process. However, in the **ORIG** and **MTV** dataset, the blocks (e.g., regions inside a large object) that receive no fixations cannot be simply treated as negative samples but just as unlabeled data. In this case, the problem turns to "learning from positive and unlabeled data." This presents a difficulty for existing learning-based algorithms since most of them are designed to learn from positive and negative data (e.g.,Kienzle et al, 2007b; Navalpakkam and Itti, 2007; Peters and Itti, 2007).

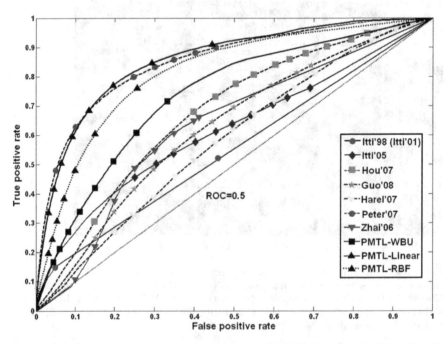

Fig. 5.10 ROC curves of different algorithms on the **RSD** benchmark. Reproduced from (Li et al, 2010), with kind permission from Springer Science+Business Media.

5.3   Knowledge-Based Saliency Model

Compared with the task-driven approaches, some learning-based approaches proposed to learn the prior knowledge that is irrelevant with specific tasks (e.g., the knowledge that helps us to process various parts of a complex object as a whole). Usually, it is believed that such priors correspond to the *mandatory top-down process* in our vision system. In most existing approaches, such top-down prior knowledge works as a "stimulus-saliency" function to select, re-weight and integrate the input visual stimuli in the feature selection stage. For example, Itti and Koch (2001b) proposed a supervised approach to learn the optimal weights for feature combination. Navalpakkam and Itti (2007) modeled the top-down gain optimization as maximizing the signal-to-noise ratio (SNR). That is, they learned linear weights for feature combination by maximizing the ratio between target saliency and distractor saliency.

Besides learning the linear feature weights, Kienzle et al (2007b) proposed a nonparametric approach to learn a visual saliency model from human eye-fixations on images. A support vector machine (SVM) was trained to determine the saliency using the local intensities. Judd et al (2009) introduced multiple low-level, mid-level and high-level features. From such a large feature pool, a linear SVM was trained to combine these features and compute the saliency value. Peters and Itti (2007) presented an approach to learn the projection matrix between global scene characteristics and eye density maps. Lu et al (2012) collected a large number of visual features, including local energy, saliency values of existing bottom-up models, car and pedestrian detectors, face detectors, convexity maps. These features were then used to train a context-aware model for image saliency computation.Zhao and Koch (2012) proposed a boosting approach to iteratively train weak classifiers from a large feature pool. These weak classifiers are then fused with linear weights to generate a visual saliency model. For video, Kienzle et al (2007a) presented an approach to learn a set of temporal filters from eye-fixations to find the interesting locations. On the regional saliency dataset, Liu et al (2007) proposed a set of novel features and adopted a Conditional Random Field (CRF) to combine these features for salient object detection. After that, they extended the approach to detect salient objects in video (Liu et al, 2008).

Generally speaking, the learning-based approaches discussed above can demonstrate promising performance on small benchmarks. The parameters trained and fine-tuned on part of the benchmark can usually achieve high performance on the rest of the benchmark. However, the most severe drawback of these approaches is that they require the supervised learning process. In this process, all the training data should be labeled with eye tracking devices or manual labeling activities, which is really time-consuming. Consequently, existing benchmarks are usually very small (with only hundreds or thousands of images), which is far from sufficient to cover all possible cases. Therefore, the trained models often have the over-fitting risk. For instance, the model trained on a limited number of images can bias to specific feature channels and locations to generate promising results on similar testing scenes, but such model may fail when encountering unknown scenes. That also hampers the further usage of these learning-based models in actual applications.

To avoid the over-fitting risk, we present a novel Bayesian approach in this Section for image saliency estimation by jointly capturing the influences of the input visual stimuli and the prior knowledge that are unsupervisedly learned from millions of images. As shown in Fig. 5.11, this approach first computes the bottom-up saliency by only considering the unbiased competitions between visual signals. After that, the bottom-up saliency is modulated by the prior knowledge statistically learned from millions of images. In particular, the foreground prior is learned by inferring the spatial distributions of all kinds of image patches, and can be used to identify whether an image patch belongs to foreground. The correlation prior is learned by mining the patch co-occurrence characteristics, which can be used to model the mutual influence between different image patches. These two priors are then used to bias the competition between visual signals by recovering the wrongly suppressed targets and inhibiting the improperly popped-out distractors. Finally, the estimated saliency maps can be improved by simultaneously using the cues from the visual signal and the prior knowledge.

The main contributions of this approach are summarized as follows:

1. Two kinds of prior knowledge, including the foreground prior and the correlation prior, are presented for estimating saliency in the free-viewing scenario. By modeling both the spatial distributions and correlations of various visual stimuli, such priors can well adapt to various scenes in recovering the wrongly suppressed targets and inhibiting the improperly popped-out distractors.
2. We propose an effective learning algorithm to learn the prior knowledge from millions of images in an unsupervised manner. Such prior knowledge, which is learned from huge amounts of images, is statistically significant and avoids the over-fitting risk.
3. A Bayesian framework is proposed to jointly capture the influences of the visual stimuli and the prior knowledge for visual saliency estimation. Experimental results show that this framework is effective to modulate any kinds of bottom-up saliency to better predict human fixations.

In the rest of this Section, we will first present the Bayesian formulation to integrate the bottom-up and top-down components. Different from the probabilistic framework in previous Section, this formulation assumes that top-down modulation occurs after the bottom-up competition. With the Bayesian framework, we describe the details on how to statistically learn the priors from millions of images and how to modulate the bottom-up saliency with the learned priors. Finally, the effectiveness of this approach is validated through extensive experiments.

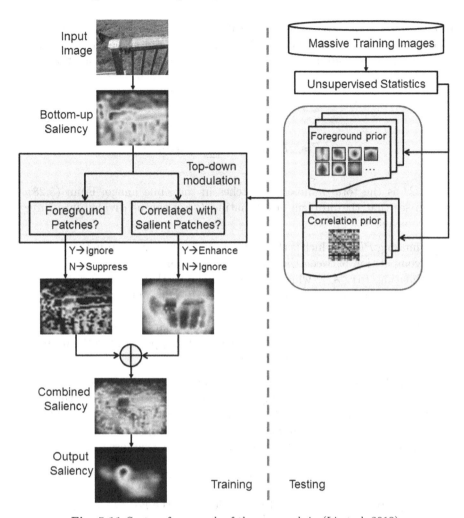

Fig. 5.11 System framework of the approach in (Li et al, 2013)

5.3.1   The Bayesian Formulation

To estimate visual saliency, one of the most important problem is to simultaneously model the influences of the visual stimuli and the prior knowledge. In human vision system, various visual stimuli will compete to become salient, while the prior knowledge may bias the competition in two ways: recovering the foreground targets that are wrongly suppressed and inhibiting the background distractors that are improperly popped-out. Following this idea, we propose a Bayesian framework to jointly capture the influences of the visual stimuli and the prior knowledge. In this framework, we focus on modulating the visual saliency acquired through bottom-up competition with various top-down priors. Let $\mathbf{s}_n$ be the event that an image patch $\mathcal{B}_n$ (e.g., 8×8 macro blocks) pops-out after the bottom-up competition and $\mathbf{r}_n$ be the event that $\mathcal{B}_n$ becomes salient after the top-down modulation, we can assume that:

$$
\begin{aligned}
P(\mathbf{r}_n) &= \sum_{m=1}^{M} P(\mathbf{s}_m)P(\mathbf{r}_n|\mathbf{s}_m) \\
&= P(\mathbf{s}_n)P(\mathbf{r}_n|\mathbf{s}_n) + \sum_{m \neq n}^{M} P(\mathbf{s}_m)P(\mathbf{r}_n|\mathbf{s}_m).
\end{aligned}
\tag{5.28}
$$

where M is the total number of patches in the same image. From (5.28), we can see that the problem of estimating $P(\mathbf{r}_n)$ can be divided into three sub-problems:

1. Estimating $P(\mathbf{s}_n)$, which is the probability that $\mathcal{B}_n$ pops-out in the bottom-up competition between the input visual stimuli;
2. Estimating $P(\mathbf{r}_n|\mathbf{s}_n)$, which is the probability that $\mathcal{B}_n$ becomes salient after the top-down modulation if it already pops-out in the bottom-up competition (as shown in Fig. 5.12(a)); and
3. Estimating $P(\mathbf{r}_n|\mathbf{s}_m)$, which is the probability that $\mathcal{B}_n$ becomes salient after the top-down modulation if another patch $\mathcal{B}_m$ pops-out in the bottom-up competition (as shown in Fig. 5.12(b)).

Among all these three sub-problems, the first one has been well studied in the past decades and there already exist many feasible solutions to this sub-problem (e.g., Itti et al (1998); Bruce and Tsotsos (2005); Harel et al (2007); Hou and Zhang (2007); Parikh et al (2008)). Here we denote the estimated bottom-up saliency as $S_b(n)$ for a patch $\mathcal{B}_n$ and thus can assume:

$$
P(\mathbf{s}_n) \propto S_b(n).
\tag{5.29}
$$

In the following study, we will mainly focus on the last two sub-problems and the main difficulty is to modulate bottom-up saliency using various kinds of prior knowledge. In this process, only the appearances and locations of the patches are available. Therefore, **the prior knowledge related to visual attributes and positional information** could probably be an effective key to solve the proposed two sub-problems.

5.3.2 Saliency with Statistical Priors

In this Section, we will address the proposed two sub-problems and modulate the bottom-up saliency with the learned top-down priors. First, we investigate what kinds of prior knowledge should be learned. After that, we describe the details on how to learn the required prior knowledge through massive image

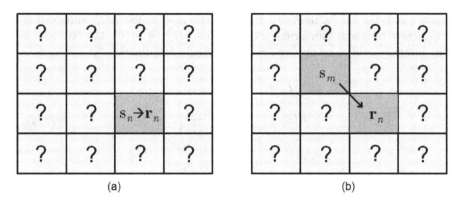

Fig. 5.12 Problem statement when modulating bottom-up saliency with top-down priors. (a) whether $\mathcal{B}_n$ can become salient after the top-down modulation when only considering its bottom-up saliency; (b) whether $\mathcal{B}_n$ can become salient after the top-down modulation when only considering the bottom-up saliency of $\mathcal{B}_m$. Reprinted from (Li et al, 2013), with kind permission from Springer Science+Business Media.

statistics. Finally, we present how to estimate visual saliency using the learned priors.

5.3.2.1   What to Learn?

Generally speaking, there are numerous kinds of prior knowledge and it is impossible to learn all of them. According to the problems stated in Fig. 5.12, we have to learn the prior knowledge that demonstrates a biased selectivity on the visual attributes and positional information, which are the only cues in conducting the top-down modulation. Therefore, we will mainly focus on two kinds of prior knowledge, including the **foreground prior** and the **correlation prior**. The foreground prior aims to identify whether an image patch belongs to foreground using its visual attributes and positional information. This prior knowledge can be helpful to estimate $P(\mathbf{r}_n|\mathbf{s}_n)$. The correlation prior aims to model the mutual correlations between image patches. This prior knowledge can be helpful to estimate $P(\mathbf{r}_n|\mathbf{s}_m)$ by taking the correlation between image patches into account. With these two kinds of prior knowledge, bottom-up saliency can be modulated to recover the wrongly suppressed targets and inhibit the improperly popped-out distractors.

In order to learn such prior knowledge and ensure the learned priors are statistically significant, we collect 9.6 million images that are randomly crawled from Flickr. Each image is resized to have a max side length of no more than 320 pixels while keeping the width-to-height ratio. Each image is then divided into a set of non-overlapping 8×8 patches, and each patch is characterized by its preattentive features (Wolfe, 2005). That is, we represent a pixel v in an image patch $\mathcal{B}_n$ by its intensity I_v, red-green opponency RG_v, blue-yellow opponency BY_v and four orientation features $\{O_v^\theta\}, \theta \in \{0°, 45°, 90°, 135°\}$. The intensity and color opponencies can be calculated as:

$$I_v = \frac{r_v + g_v + b_v}{3},$$

$$RG_v = \frac{r_v - g_v}{\max(r_v, g_v, b_v)}, \qquad (5.30)$$

$$BY_v = \frac{b_v - \min(r_v, g_v)}{\max(r_v, g_v, b_v)},$$

where r_v, g_v, b_v are the red, green and blue components for pixel v, respectively. Here the red-green and blue-yellow opponencies are calculated as in (Walther and Koch, 2006), which will be set to zero if $\max(r_v, g_v, b_v) < 0.1$ to avoid large fluctuations at low luminance.

The orientation feature O_v^θ can be calculated by convolving I_v with Gabor filters:

$$O_v^\theta = \| I_v * G_0(\theta) \| + \| I_v * G_{\pi/2}(\theta) \|, \qquad (5.31)$$

where $G_0(\theta)$ and $G_{\pi/2}(\theta)$ are two Gabor filters oriented at θ with phase 0 and $\pi/2$, respectively[2]. After calculating these features for all pixels in $\mathcal{B}_n$, we further quantize each feature into 4 bins to acquire 7 histograms from an image patch $\mathcal{B}_n$. Finally, each patch is characterized by a feature vector with 7×4=28 components with the same dynamic range of [0,1].

From 9.6 million images, we can obtain billions of image patches (i.e., billions of 28d feature vectors). To efficiently learn prior knowledge from such a huge number of image patches, we have to further reduce the feature dimension to obtain a more compact patch representation. Toward this end, a feasible solution could be generating a set of visual words by performing k-means clustering on all the image patches and represent each patch with the nearest visual word. However, it is very difficult to directly perform the k-means clustering on billions of image patches due to the limitation of computational resource. Therefore, we use the affinity propagation algorithm proposed in (Frey and Dueck, 2007) to select a set of representative patches (i.e., exemplars) from each image. In this process, the number of exemplars is automatically determined according to the complexity of image content. Since such exemplars are much fewer than the original patches, we can perform k-means clustering on their feature vectors to form a vocabulary of N_w visual words, denoted as $\{\mathbf{w}_i\}_{i=1}^{N_w}$ (in the experiments, we will show the influence of N_w). With these visual words, each patch can be quantized to the nearest visual word using the Euclidian distance measure. Finally, we can represent a patch $\mathcal{B}_n$ with only one integer label $l_n \in \{1, ..., N_w\}$.

---

[2] For more explanations about extracting these preattentive features, please refer to Sect. 3.2.1

5.3.2.2   Learning the Foreground Prior

To calculate the foreground prior, we simply count the times that a visual word $\mathbf{w}_i$ appears at any probable locations. After that, we can get N_w distribution maps $\{\mathcal{D}_i\}_{i=1}^{N_w}$, which are then normalized to let the location with the highest frequency corresponds to 1. Some representative visual words and their distribution maps are shown in Fig. 5.13.

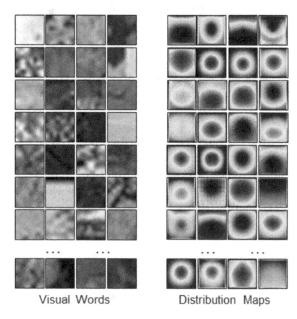

Fig. 5.13 Representative visual words and their distribution maps. Reprinted from (Li et al, 2013), with kind permission from Springer Science+Business Media.

Visual Words   Distribution Maps

From Fig. 5.13, we can see that many visual words demonstrate center-biased distributions, while several visual words distribute around image edges. Since people often snap photos by intentionally placing the targets near to image centers (i.e., the photographer bias in Tseng et al (2009)), foreground targets often appear around image centers while background distractors usually appear near to image edges (as shown in Fig. 5.14). Therefore, we can safely assume that **visual words have higher probabilities to appear in the foreground than in the background if they distribute around image centers**.

Following this assumption, we can use the distribution maps as the foreground criterion. As shown in Fig. 5.15(a), we divide a distribution map $\mathcal{D}_i$ into two regions with equivalent area and use Ω_i to quantify its center-bias property as the percentage of energy in the "center" region. From Fig. 5.15(b), we can see that the quantified center-bias properties are high on many distribution maps, while some maps are obviously edge-biased. Thus the corresponding patches can be treated as distractors and should be suppressed.

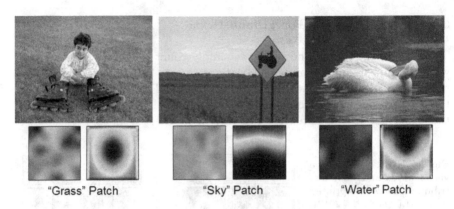

"Grass" Patch   "Sky" Patch   "Water" Patch

Fig. 5.14 The distribution maps of typical background patches. Reprinted from (Li et al, 2013), with kind permission from Springer Science+Business Media.

Given the quantified center-bias property of each visual word, we can infer the probability that an image patch belongs to foreground using its visual attributes and positional information. Suppose that $\mathcal{B}_n$ is classified to the visual word $\mathbf{w}_{l_n}$ and let $\mathbf{f}_n$ be the event that $\mathcal{B}_n$ is a foreground patch, we can estimate $P(\mathbf{f}_n)$ as:

$$P(\mathbf{f}_n) \propto [\Omega_{l_n} \geq 0.5]_\mathbf{I} \cdot \mathcal{D}_{l_n}(n), \tag{5.32}$$

where $[\Omega_{l_n} \geq 0.5]_\mathbf{I}$ equals to 1 if $\Omega_{l_n} \geq 0.5$ and 0 otherwise. $\mathcal{D}_{l_n}(n)$ indicates the frequency that the visual word $\mathbf{w}_{l_n}$ appears at the location of $\mathcal{B}_n$. From (5.32), we can see that the probability that $\mathcal{B}_n$ belongs to foreground could become high if: 1) $\mathbf{w}_{l_n}$ distributes around image centers (i.e., $\Omega_{l_n} \geq 0.5$) and thus has a higher probability to appear in the foreground than in the background; and 2) $\mathcal{B}_n$ appears at recurring locations that a probable foreground visual word $\mathbf{w}_{l_n}$ can be frequently observed in millions of images.

5.3.2.3   Learning the Correlation Prior

The objective of learning the correlation prior is to model the mutual influence between any two patches. Generally speaking, there exist two kinds of typical mutual influences: 1) if two image patches $\mathcal{B}_m$ and $\mathcal{B}_n$ are correlated, they will probably excite each other. Once we observe one patch, we may expect the other one; 2) on the contrary, irrelevant image patches will compete to inhibit each other. Here we use Υ_{mn} to quantify the correlation strength between two visual words $\mathbf{w}_m$ and $\mathbf{w}_n$. In calculating Υ_{mn}, two visual words that frequently co-occur in the same images may be tightly correlated. That is, if we can observe one visual word, we can expect another one with a high probability. Moreover, the probability of expecting $\mathbf{w}_n$ when $\mathbf{w}_m$ is observed

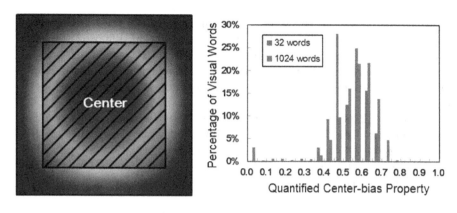

Fig. 5.15 Quantified center-biased property of distribution maps. Reprinted from (Li et al, 2013), with kind permission from Springer Science+Business Media.

should be different from that of expecting $\mathbf{w}_m$ when $\mathbf{w}_n$ is observed. For example, many images may contain a common visual word $\mathbf{w}_n$. When some other visual words in these images are observed, $\mathbf{w}_n$ can be expected with high probabilities. However, we can hardly expect other specific visual words when $\mathbf{w}_n$ is observed.

Following this idea, we can estimate Υ_{mn} using massive image statistics. First, we count the frequency F_n which indicates the total times that the visual word $\mathbf{w}_n$ appears in all the training images. Meanwhile, we also count the frequency F_{mn} which represents the total times that two visual words $\mathbf{w}_m$ and $\mathbf{w}_n$ appear in the same images (note that $F_{nn}=F_n$). After that, we can calculate Υ_{mn} as:

$$\Upsilon_{mn} = \frac{F_{mn}}{F_n}. \tag{5.33}$$

Fig. 5.16 The histogram of quantified correlations $\{\Upsilon_{mn}\}$ between 1024 visual words. Reprinted from (Li et al, 2013), with kind permission from Springer Science+Business Media.

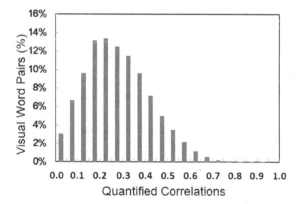

From (5.33), we can see that Υ_{mn} is unequal to Υ_{nm} and a higher co-occurrence frequency will lead to a stronger correlation. Figure 5.16 shows the histogram of such quantified correlation strength between 1024 visual words. We can see that most visual words have weak correlations, while some visual words demonstrate strong co-occurrence properties even in millions of images.

Given the quantified correlation strength between visual words, we can infer the probability that one image patch is tightly correlated with another patch using their visual attributes and positional information. Suppose that $\mathcal{B}_m$ and $\mathcal{B}_n$ are classified to visual words $\mathbf{w}_{l_m}$ and $\mathbf{w}_{l_n}$ and let $\mathbf{o}_{mn}$ be the event that $\mathcal{B}_n$ is correlated to $\mathcal{B}_m$, we can estimate $P(\mathbf{o}_{mn})$ as:

$$P(\mathbf{o}_{mn}) \propto \Upsilon_{l_m l_n} \cdot \mathcal{N}(d_{mn}; 0, \sigma_c), \tag{5.34}$$

where $\mathcal{N}$ is the Gaussian distribution and σ_c is empirically set to 0.3 in this study. d_{mn} is the distance between $\mathcal{B}_m$ and $\mathcal{B}_n$, which is normalized by the distance from image corner to image center. The Gaussian term is important to ensure that only the correlations between the patches in a local area are considered to increase the computational efficiency. From (5.34), we can see that the probability that $\mathcal{B}_n$ is correlated to $\mathcal{B}_m$ will be high if: 1) $\mathbf{w}_{l_m}$ and $\mathbf{w}_{l_n}$ frequently co-occur in millions of images; and 2) $\mathcal{B}_m$ and $\mathcal{B}_n$ are near to each other.

5.3.2.4   Computing Visual Saliency with Statistical Priors

Given the learned foreground prior and correlation prior, we can now turn to the two sub-problems proposed above: how to estimate $P(\mathbf{r}_n|\mathbf{s}_n)$ and $P(\mathbf{r}_n|\mathbf{s}_m)$?

To estimate $P(\mathbf{r}_n|\mathbf{s}_n)$, we have to first infer the foreground prior $P(\mathbf{f}_n)$ using (5.32) to see whether $\mathcal{B}_n$ is a foreground patch. With the foreground prior, we can rewrite $P(\mathbf{r}_n|\mathbf{s}_n)$ as:

$$P(\mathbf{r}_n|\mathbf{s}_n) = P(\mathbf{f}_n)P(\mathbf{r}_n|\mathbf{s}_n, \mathbf{f}_n) + P(\bar{\mathbf{f}}_n)P(\mathbf{r}_n|\mathbf{s}_n, \bar{\mathbf{f}}_n). \tag{5.35}$$

From (5.35), we can see that there are two probable combinations of events $\mathbf{s}_n$ and $\mathbf{f}_n$, including:

- $\mathbf{s}_n$ and $\mathbf{f}_n$: $\mathcal{B}_n$ is a target that is correctly popped-out by the bottom-up model.
- $\mathbf{s}_n$ and $\bar{\mathbf{f}}_n$: $\mathcal{B}_n$ is probably a distractor that is improperly popped-out by the bottom-up model.

When modulating the bottom-up saliency with the foreground prior, we can maintain the correctly popped-out targets and suppress the improperly popped-out distractors by setting:

$$P(\mathbf{r}_n|\mathbf{s}_n, \mathbf{f}_n) \approx 1, \ P(\mathbf{r}_n|\mathbf{s}_n, \bar{\mathbf{f}}_n) = e^{-\alpha_b}. \tag{5.36}$$

where $\alpha_b \geq 0$ is a predefined constant to fuse the conflict predictions made by the bottom-up saliency model and the foreground prior. Smaller $e^{-\alpha_b}$ indicates the foreground prior is more reliable (we will show the influence of α_b in the experiment). By incorporating (5.36) into (5.35), we can estimate $P(\mathbf{r}_n|\mathbf{s}_n)$ as:

$$P(\mathbf{r}_n|\mathbf{s}_n) = e^{-\alpha_b} + (1 - e^{-\alpha_b})P(\mathbf{f}_n), \tag{5.37}$$

where the foreground prior $P(\mathbf{f}_n)$ can be estimated using (5.32). From (5.37), we can see that the bottom-up saliency can be selectively modulated by the foreground prior. In this process, the real targets, which are predicted as salient by both the bottom-up saliency model and foreground prior, will become salient. On the contrary, the distractors, which pop-out in the bottom-up competition, will be suppressed if the foreground prior classifies them as distractors.

To estimate $P(\mathbf{r}_n|\mathbf{s}_m)$, we have to first infer the correlation prior $P(\mathbf{o}_{mn})$ to find whether $\mathcal{B}_n$ is tightly correlated with $\mathcal{B}_m$. Inspired by this idea, we have:

$$\begin{aligned} P(\mathbf{r}_n|\mathbf{s}_m) =& P(\mathbf{o}_{mn})P(\mathbf{r}_n|\mathbf{s}_m, \mathbf{o}_{mn}) \\ &+ P(\bar{\mathbf{o}}_{mn})P(\mathbf{r}_n|\mathbf{s}_m, \bar{\mathbf{o}}_{mn}). \end{aligned} \tag{5.38}$$

From (5.38), we can also find two probable combinations of events $\mathbf{s}_n$ and $\mathbf{o}_{mn}$, including:

- $\mathbf{s}_m$ and $\mathbf{o}_{mn}$: $\mathcal{B}_n$ is tightly correlated with a patch that pops-out in the bottom-up competition. In this case, $\mathcal{B}_n$ will be excited by $\mathcal{B}_m$.
- $\mathbf{s}_m$ and $\bar{\mathbf{o}}_{mn}$: $\mathcal{B}_n$ is irrelevant with a patch that pops-out in the bottom-up competition. In this case, $\mathcal{B}_n$ will be inhibited by $\mathcal{B}_m$.

When modulating the bottom-up saliency with the correlation prior, two tightly correlated patches will excite each other while irrelevant patches will inhibit each other by setting:

$$P(\mathbf{r}_n|\mathbf{s}_m, \mathbf{o}_{mn}) \approx 1, \ P(\mathbf{r}_n|\mathbf{s}_m, \bar{\mathbf{o}}_{mn}) \approx 0. \tag{5.39}$$

By incorporating (5.39) into (5.38), we can thus estimate $P(\mathbf{r}_n|\mathbf{s}_m)$ as:

$$P(\mathbf{r}_n|\mathbf{s}_m) = P(\mathbf{o}_{mn}), \tag{5.40}$$

where the correlation prior $P(\mathbf{o}_{mn})$ can be estimated using (5.34). From (5.40), we can see that the wrongly suppressed targets may become salient after the top-down modulation if it is tightly correlated with the targets that pop-out in the bottom-up competition. As shown in Fig. 5.17, the bottom-up saliency map of an image, which is calculated by (Itti et al, 1998), sometimes only pops-out the borders of the large salient target (e.g., the head and tail of the cow), while the inner smooth parts are ignored (e.g., the body of

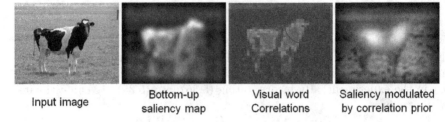

| Input image | Bottom-up saliency map | Visual word Correlations | Saliency modulated by correlation prior |

Fig. 5.17 The correlation prior can help to recover the wrongly suppressed targets and to pop-out large salient target as a whole. Reprinted from (Li et al, 2013), with kind permission from Springer Science+Business Media.

the cow). To recover the wrongly suppressed targets, we first estimate the correlation strength between visual words. Figure 5.17 shows the correlation strength Υ_{mn} between the patch marked in red and all the other patches. To increase the computational efficiency, we only consider the correlations between nearby patches (e.g., patches in the red circle controlled by a Gaussian term). Finally, the border patch marked with red, which successfully pops-out in the bottom-up competition, will help to recover the wrongly suppressed targets (i.e., large $P(\mathbf{s}_m)$ and $P(\mathbf{r}_n|\mathbf{s}_m)$ will lead to large $P(\mathbf{r}_n)$). In this manner, we can pop-out the salient target as a whole (as shown in Fig. 5.17), especially for those objects with large smooth regions.

After estimating $P(\mathbf{r}_n|\mathbf{s}_n)$ and $P(\mathbf{r}_n|\mathbf{s}_m)$, the saliency value of $\mathcal{B}_n$ after the top-down modulation, denoted as $S_r(n) \propto P(\mathbf{r}_n)$, can thus be calculated by incorporating (5.29), (5.37) and (5.40) into (5.28):

$$
\begin{aligned}
S_r(n) \propto\ & S_b(n) \cdot (e^{-\alpha_b} + (1 - e^{-\alpha_b})P(\mathbf{f}_n)) \\
& + \sum_{m \neq n}^{M} S_b(m) \cdot P(\mathbf{o}_{mn}),
\end{aligned}
\tag{5.41}
$$

where $P(\mathbf{f}_n)$ and $P(\mathbf{o}_{mn})$ are foreground and correlation priors that can be derived using (5.32) and (5.34), respectively. To facilitate the computation of $S_r(n)$, we first obtain two saliency maps using the first and the second terms in (5.41), respectively. These two saliency maps are then normalized into the same range of [0,1] and fused with equal weights.

Given a testing image, its saliency map can be easily estimated through four major steps (as shown in Fig. 5.11):

1. Resize the image to have a max side length of no more than 320 pixels and divided the image into non-overlapping 8×8 patches;
2. Use any existing bottom-up model to estimate a bottom-up saliency value for each patch;
3. Estimate the foreground and correlation priors using (5.32) and (5.34) and then use them to generate the modulated saliency map using (5.41); and

4. Convolve the saliency map with a disk filter (with the radius of 3) to fill
   in the "holes" generated by applying inconsistent foreground priors on
   adjacent patches in smooth regions. Then conduct an exponential operation $S_r^*(n)=S_r(n)^3$ to remove the fuzzy background generated by using
   the additive Bayesian formulation.

From these processes, we can see that our proposed approach is biologically
plausible since neurobiological evidences show that the bottom-up factors in
human vision system act faster than the top-down factors (Wolfe et al, 2000;
Henderson, 2003). Visual signals will first compete fairly to generate the
bottom-up saliency, while a slower recall or recognition process is conducted
to load the related prior knowledge into the working memory to bias the
competition. In this process, the bottom-up saliency maps are modulated by
various top-down priors to pop-out the real targets and suppress the real
distractors. Moreover, we can also see that our approach exhibits good generalization abilities and can be easily extended. On one hand, we can plug
any state-of-the-art bottom-up saliency model into the proposed framework,
no matter how it detects saliency. On the other hand, if we can learn more
kinds of prior knowledge (e.g., the task-dependent priors), we can easily incorporate them into our framework by calculating more kinds of top-down
saliency maps, leading to a more accurate estimation of visual saliency.

5.3.3 Experiments

In this section, several experiments are conducted to prove the effectiveness
of our approach. The main objectives are two folds: 1) to evaluate whether
the prior knowledge is useful in estimating visual saliency and 2) to explore
how the prior knowledge works in the estimation processes. Toward this end,
we adopt two datasets in the experiments, including:

- **Toronto-120**. This popular dataset was proposed in (Bruce and Tsotsos,
  2005) and has been used in many recent studies on visual saliency. It
  contains 120 color images. On each image, the fixations from 20 different
  subjects were recorded under the free-viewing conditions to reveal the
  locations of the salient targets.
- **MIT-1003**. This dataset was provided by Judd et al (2009). It consists of
  1,003 images in total, most of which are color images. The eye tracing data
  were recorded from 15 subjects who free viewed these images. Compared
  with **Toronto-120**, this dataset is more challenging since images in this
  dataset are usually more complex and most of them contain a lot of targets
  and distractors.

On these two datasets, we adopt 10 approaches for comparison. All the
source codes or executables can be found on the Internet. These approaches
can be roughly categorized into two groups, including:

- **BU Group**. This group contains six bottom-up approaches, including Itti98[3] (Itti et al, 1998), Harel07 (Harel et al, 2007), Hou07 (Hou and Zhang, 2007), Achanta09 (Achanta et al, 2009), Goferman10[4](Goferman et al, 2010) and Riche12 (Riche et al, 2012). These bottom-up approaches only utilize the input visual signals to generate the bottom-up saliency maps. By comparing our approach with them, we wish to prove that incorporating the learned statistical priors can improve the performance of visual saliency estimation by modulating bottom-up saliency.
- **STAT Group**. This group contains four statistical approaches, including Bruce05 (Bruce and Tsotsos, 2005), Hou09 (Hou and Zhang, 2009), Zhang08 (Zhang et al, 2008) and Wang10 (Wang et al, 2010). These approaches also utilize the statistical image priors. By comparing our approach with them, we wish to prove that our framework is more effective in utilizing the learned prior knowledge.

In the comparison, we use the Area Under the ROC Curve (**AUC**) for performance evaluation. In particular, we re-sample the non-fixated pixels by using the approach described in Chap. 2.3.1 to make fair comparisons. Moreover, there are usually two ways to evaluate the overall performance on multiple images: 1) calculate the **AUC** score on each image first and then compute the mean and standard deviation of all the **AUC** scores; and 2) summing up the numbers of true positives, true negatives, false positives and false negatives on all images and generate a unique ROC curve, leading to a unique **AUC** score. Both ways can make sense and we will adopt the first way in the following experiments.

5.3.3.1   Whether It Works

In the first experiment, the main objective is to see whether our approach can really work. Toward this end, we adopt 6 bottom-up models to see whether the prior knowledge learned by our approach is effective to modulate the bottom-up saliency. In this process, we use 32 visual words and set $e^{-\alpha_b} \approx 0$ (the influences of these parameters will be discussed in other experiments). We also compare the modulated saliency maps with those maps generated by 4 approaches in the statistical group to see whether our framework can utilize the learned prior knowledge in a more effective manner. The **AUC** scores are shown in Table. 5.5. Some representative examples are illustrated in Fig. 5.18. Note that the fixation density maps are generated by filtering the pixel-wise fixation maps using a Gaussian kernel to account for inaccurate tracking results and the decreasing visual accuracy with increasing eccentricity from the fovea.

---

[3] The "winner-take-all" competition is not used in Itti98.

[4] The face detection component is not activated in Goferman10.

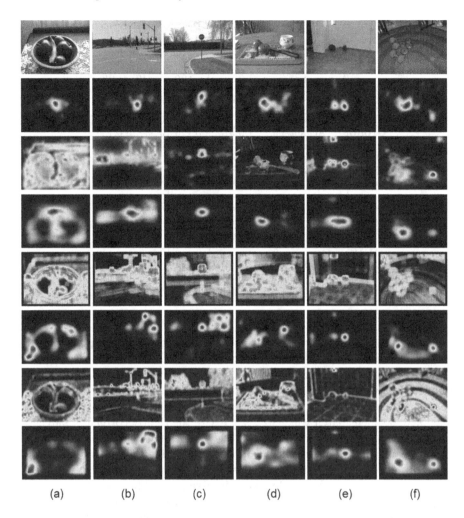

(a)   (b)   (c)   (d)   (e)   (f)

Fig. 5.18 Some representative saliency maps generated by various saliency models. The first row shows the input images and the second row illustrates the corresponding fixation density maps. The third row contains bottom-up saliency maps calculated by (a) Itti98; (b) Harel07; (c) Hou07; (d) Achanta09; (e) Goferman10; (f) Riche12. The 4th row demonstrates our results acquired by modulating the bottom-up saliency maps with the learned priors. The last four rows are results from Bruce05, Hou09, Zhang08 and Wang10, respectively. Reprinted from (Li et al, 2013), with kind permission from Springer Science+Business Media.

Table 5.5 Performance of various approaches on the two image benchmarks. Reproduced from (Li et al, 2013), with kind permission from Springer Science+Business Media.

	Approaches	Toronto-120	MIT-1003
BU Group	Itti98	0.731 ± 0.123	0.678 ± 0.134
	Harel07	0.762 ± 0.134	0.700 ± 0.152
	Hou07	0.763 ± 0.122	0.693 ± 0.142
	Achanta09	0.575 ± 0.126	0.554 ± 0.130
	Goferman10	0.797 ± 0.100	0.713 ± 0.140
	Riche12	0.821 ± 0.090	0.722 ± 0.135
STAT Group	Bruce05	0.758 ± 0.109	0.700 ± 0.123
	Hou09	0.787 ± 0.112	0.708 ± 0.153
	Zhang08	0.705 ± 0.129	0.667 ± 0.136
	Wang10	0.786 ± 0.113	0.704 ± 0.152
OUR Group	**Our+Itti98**	0.794 ± 0.118	0.714 ± 0.143
	Our+Harel07	0.804 ± 0.122	0.710 ± 0.153
	Our+Hou07	0.797 ± 0.112	0.706 ± 0.148
	Our+Achanta09	0.710 ± 0.148	0.637 ± 0.168
	Our+Goferman10	0.816 ± 0.102	0.725 ± 0.140
	Our+Riche12	$\mathbf{0.834 \pm 0.086}$	$\mathbf{0.738 \pm 0.13}$

From Table. 5.5, we can see that the priors learned by our approach can improve the saliency maps generated by all the 6 bottom-up approaches. No matter how the bottom-up competitions are conducted in these approaches, our learned prior knowledge can effectively recover the wrongly suppressed targets and inhibit the improperly popped-out distractors. As shown in Fig. 5.18, a salient patch will *selectively* excite the tightly correlated patches using the correlation prior, while the distractors, especially the common background patches, can be effectively suppressed by using the foreground prior. In traditional bottom-up models, high saliency values are usually assigned to unique or rare visual subsets. However, the assumption that visual rarity corresponds to high saliency may not always hold since the background patches can sometimes become unique or rare. These patches, which are already very familiar to the subjects, will be easily ignored. However, the bottom-up approaches will equally treat all the input signals since they have no prior knowledge on what the patch is. In our approach, we find that such common distractors often distribute around image edges. Therefore, we learn the distribution maps and quantify their center-bias properties to determine whether a patch is a common background patch or not. Then these patches will be effectively recognized and suppressed.

From Table 5.5, we can also see that the modulated saliency maps from Itti98, Harel07, Hou07, Goferman10 and Riche12 can better predict human fixations than those saliency maps generated by another four approaches in the statistical group. This is mainly due to two reasons. First, most of the parameters used in our approach (e.g., the visual words, foreground and correlation priors) are learned from millions of images. Therefore, our approach can well handle the outliers. Second, we have adopted an opposite way to use the learned priors. In our approach, each patch is quantified to the nearest visual word and represented only by an integer label. In this process, many details are discarded but the integer label can work well since its main role is to retrieve the related prior knowledge. On the contrary, the other approaches will map the patch into a subspace with much higher dimensions and then estimate visual saliency in that subspace. Since the subspace may be not optimal, there may generate rich redundancies in the mapping. As illustrated in Fig. 5.18, these redundancies may generate many "noise" in the estimated saliency maps since it can be very difficult to distinguish targets from distractors when projecting all the signals onto specific basis. Therefore, these approaches achieve lower **AUC** scores.

Generally speaking, the main difference between our approach and all the other approaches discussed above lies in that we treat the input signals with bias. That is, each kind of prior knowledge will demonstrate a specific kind of biased selectivity in visual saliency estimation. For instance, the foreground prior will selectively suppress the patches that are judged as distractors, while the correlation prior will selectively enhance the patches around existing salient patches. Actually, such selectivity is well supported by biological evidences, which have proved that the top-down factors can bias the competition between the neurons linked with visual stimuli by favoring a specific category of stimuli.

In particular, the priors used in our approach are statistically learned from massive images in an unsupervised manner. Compared with (Torralba et al, 2006) and (Chikkerur et al, 2010) that mainly focused on incorporating the task-dependent priors, our approach can be used in much more scenarios to predict human fixations under free-viewing conditions since we have no assumption on the probable target-of-interests. Another advantage of learning the prior knowledge from millions of images is that the over-fitting risk can be largely avoided. Compared with the models trained on hundreds of images, our model often demonstrates impressive generalization ability. For instance, Judd et al (2009) selected 903 images from **MIT-1003** and extracted a set of low-, mid- and high-level features as well as the center prior to train a linear SVM model as the saliency model. Even with such a large feature pool, the **AUC** only reached 0.725 on the rest 100 images. Actually, if we adopt the same center-surround contrast features used in Itti98 to train the linear SVM model, the **AUC** will decrease to 0.684. This is natural since simple linear weights often lack the ability to model complex priors. Actually, people may attend to the salient targets in limited training images by only focusing on some specific features (e.g., human face as a special case). However, these features, which can be mined through supervised learning algorithms, may not always work well on the testing images (i.e., over-fitting). Therefore, it is necessary to learn the prior knowledge from massive unlabeled training images, probably by using unsupervised learning algorithms.

5.3.3.2   How It Works

To further investigate how the learned priors work in the top-down modulation, we conduct several experiments on the **Toronto-120** dataset to see the
influence of various parameters and top-down priors. In these experiments,
we adopt the bottom-up saliency maps generated by Itti98, which is treated
as a baseline approach with **AUC**=0.731.

First, we conduct an experiment to see the influence of the number of
visual words. In the experiment, we test 4, 8, 16, 32, 48, 64, 128, 256, 512
and 1024 visual words, and the **AUC** scores are shown in Fig. 5.19(a). From
Fig. 5.19(a), we can see that our approach performs the best when using
32 visual words. When using more visual words, the performance gradually
decreases. Although more visual words can better describe the details of
the input images, they will also become more sensitive to noise and small
fluctuations. For instance, two image patches with similar contents may be
mistakenly quantized to different visual words. Due to the probable increase
of such classification errors when using more visual words, the learned prior
knowledge will become less reliable. Moreover, when using visual words less
than 32, the influence of foreground prior will greatly decrease. For instance,
when using 4 visual words, each visual word appears at each specific location
with almost the same frequency. In this case, it is difficult to identify whether
a patch belongs to foreground or not.

Second, we conduct an experiment to see the influence of α_b (i.e., $e^{-\alpha_b} =$
$P(\mathbf{r}_n|\mathbf{s}_n, \bar{\mathbf{f}}_n)$), which indicates whether to trust the foreground prior when
it makes conflict prediction with the bottom-up saliency model. When α_b
is large (i.e., $e^{-\alpha_b}$ is small), we choose to trust the foreground prior, and
vice versa. In the experiment, we vary α_b from $+\infty$ to 0. Equivalently, $e^{-\alpha_b}$
changes from 1 to 0 and the **AUC** scores are shown in Fig. 5.19(b). From
Fig. 5.19(b), we can see that setting $e^{-\alpha_b} \approx 0$ can guarantee the best performance for **Our**+Itti98, which proves the effectiveness on the learned foreground prior.

Third, we conduct an experiment to show the influences of foreground
and correlation priors. By setting $P(\mathbf{r}_n|\mathbf{s}_m)=0$ in (5.28), we find that the
AUC can reach 0.771 when only using the foreground prior. In contrast,
the **AUC** can reach 0.746 when using only the correlation prior by setting
$P(\mathbf{r}_n|\mathbf{s}_n)=1$ in (5.28). By combining these two kinds of prior knowledge, the
overall **AUC** can reach 0.794. In particular, we find that the post-smoothing
also contributes to the overall performance, while the exponential operation,
which can often provide a "cleaner" viewing effect, has almost no influence
on the **AUC** scores since it will not change the order of patch saliency values.
When the post-smoothing operations are not used, the **AUC** can only reach
0.782. The reason is that there may exist some "holes" when modulating
the bottom-up saliency using the foreground prior since adjacent patches in
smooth regions sometimes are wrongly classified to different visual words.

Fig. 5.19 The **AUC** scores of our approach on the **Toronto-120** dataset when using different parameters such as (a) different number of visual words and (b) different α_b. Note that here the error bar corresponds to $\frac{\sigma}{\sqrt{N}}$, where σ is the standard derivation of **AUC** and N is the total number of images in **Toronto-120**. Reprinted from (Li et al, 2013), with kind permission from Springer Science+Business Media.

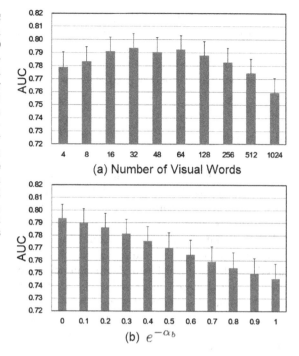

(a) Number of Visual Words

(b) $e^{-\alpha_b}$

The overall **AUC** will probably decrease without filling such "holes" using the post-smoothing operation.

To sum up, the proposed approach can work well in visual saliency estimation and demonstrate several advantages in utilizing the prior knowledge. Actually, the whole framework can be uniquely characterized by two main phases, one *fast* bottom-up phase and one *slow* top-down phase. The bottom-up phase is mainly driven by data and transfers signals in a feed-forward manner. In the transmission, certain attributes of the data will be gradually extracted to active the related prior knowledge to generate feed-backward control signals. Compared with the models that contain pure bottom-up or top-down phase or parallel bottom-up/top-down phases, such framework has been proved to be consistent with the neurobiological mechanisms demonstrated in human perception experiments and takes advantage of optimizing each phase separately (Han and Zhu, 2009).

Moreover, the framework in our approach can be easily extended. Once we learn some new kinds of prior knowledge, we can easily add them into our framework like the foreground and correlation priors. With this additive framework, we believe that the performance of visual saliency estimation can be gradually improved and a "perfect" model is expectable. Furthermore, our approach can be easily distributed on multiple computing units. This is very important since the learned knowledge database could become extremely large in the future (e.g., thousand kinds of prior knowledge). In our framework, different kinds of prior knowledge can be deployed on different computers, each of which can bias the competition of the input stimuli to generate a specific top-down saliency map and numerous top-down saliency maps can be fused to better predict human fixations.

5.4   Notes

This Chapter discusses the learning-based approaches for visual saliency computation. In particular, two novel approaches are presented to incorporate two different types of top-down factors into visual saliency computation. The first approach takes the influence of "task" into account which is considered to be related with the volitional top-down process. In this approach, we present a probabilistic multi-task learning framework to incorporate task-related top-down control into visual saliency computation. Under this framework, the task-related prior is learned to bias the feature selection in the form of "stimulus-saliency" functions and model fusion strategies. With the probabilistic framework, such scene-specific top-down control can effectively locate the salient region while suppressing the distractors.

Different from the task-related perspective, the second approach considers the influence of prior knowledge which is mainly related with the mandatory top-down process. In this approach, two kinds of priors are statistically learned from millions of images in an unsupervised manner. A Bayesian framework is then proposed to jointly take the influences of visual stimuli and prior knowledge into account. In this framework, the bottom-up saliency is calculated to pop-out the visual subsets that are probably salient, while the prior knowledge is used to recover the wrongly suppressed targets and inhibit the improperly popped-out distractors. From the experimental results, we can see that such statistical priors are highly effective in recovering the wrongly suppressed targets and removing the improperly popped-out distractors.

When validating the effectiveness of the two proposed approaches in extensive experiments, we also find two interesting phenomena. First, the proposed multi-task learning approach performs the best on advertisement and outdoor genres, while the performance on news and crowd scenes are much worse. One possible explanation is that different top-down factors should be incorporated to process scenes in different domains. Therefore, Chap. 6 will discuss how to mine the domain-specific knowledge. Second, the performance of the learning-based approaches vary remarkably on the fixation benchmark and the regional saliency benchmark. We can see that the fixation points in each video frame are very sparse, while many salient objects remain unlabeled. Therefore, Chap. 7 will explore how to improve the visual saliency model by removing label ambiguities in user data.

References

Achanta, R., Hemami, S., Estrada, F., Süsstrunk, S.: Frequency-tuned salient region detection. In: Preceedings of the IEEE Conference on Computer Vision and Pattern Recognition (CVPR), pp. 1597–1604 (2009), doi:10.1109/CVPR.2009.5206596

Argyriou, A., Evgeniou, T., Pontil, M.: Multi-task feature learning. In: Advances in Neural Information Processing Systems (NIPS), pp. 41–48 (2007)

Bruce, N.D., Tsotsos, J.K.: Saliency based on information maximization. In: Advances in Neural Information Processing Systems (NIPS), Vancouver, BC, Canada, pp. 155–162 (2005)

Cheung, C.H., Po, L.M.: A novel cross-diamond search algorithm for fast block motion estimation. IEEE Transactions on Circuits and Systems for Video Technology 12(12), 1168–1177 (2002), doi:10.1109/TCSVT.2002.806815

Chikkerur, S., Serre, T., Tan, C., Poggio, T.: What and where: A bayesian inference theory of attention. Vision Research 50(22), 2233–2247 (2010), doi:10.1016/j.visres.2010.05.013

Chun, M.M.: Contextual guidance of visual attention. In: Itti, L., Rees, G., Tsotsos, J. (eds.) Neurobiology of Attention, 1st edn., pp. 246–250. Elsevier Press, Amsterdam (2005)

Evgeniou, T., Micchelli, C.A., Pontil, M.: Learning multiple tasks with kernel methods. Journal of Machine Learning Research 6, 615–637 (2005)

Frey, B.J., Dueck, D.: Clustering by passing messages between data points. Science 315, 972–976 (2007)

Frith, C.: The top in top-down attention. In: Itti, L., Rees, G., Tsotsos, J. (eds.) Neurobiology of Attention, 1st edn., pp. 105–108. Elsevier Press, Amsterdam (2005)

Goferman, S., Zelnik-Manor, L., Tal, A.: Context-aware saliency detection. In: Preceedings of the IEEE Conference on Computer Vision and Pattern Recognition (CVPR), pp. 2376–2383 (2010), doi:10.1109/CVPR.2010.5539929

Guo, C., Ma, Q., Zhang, L.: Spatio-temporal saliency detection using phase spectrum of quaternion fourier transform. In: Preceedings of the IEEE Conference on Computer Vision and Pattern Recognition (CVPR), pp. 1–8 (2008), doi:10.1109/CVPR.2008.4587715

Han, F., Zhu, S.C.: Bottom-up/top-down image parsing with attribute grammar. IEEE Transactions on Pattern Analysis and Machine Intelligence 31(1), 59–73 (2009), doi:10.1109/TPAMI.2008.65

Harel, J., Koch, C., Perona, P.: Graph-based visual saliency. In: Advances in Neural Information Processing Systems (NIPS), pp. 545–552 (2007)

Henderson, J.M.: Human gaze control during real-world scene perception. Trends in Cognitive Sciences 7(11), 498–504 (2003)

Hou, X., Zhang, L.: Saliency detection: A spectral residual approach. In: Proceedings of the IEEE Conference on Computer Vision and Pattern Recognition (CVPR), pp. 1–8 (2007), doi:10.1109/CVPR.2007.383267

Hou, X., Zhang, L.: Dynamic visual attention: Searching for coding length increments. In: Advances in Neural Information Processing Systems (NIPS), pp. 681–688 (2009)

Itti, L.: Models of bottom-up and top-down visual attention. PhD thesis, California Institute of Technology (2000)

Itti, L.: Crcns data sharing: Eye movements during free-viewing of natural videos. In: Collaborative Research in Computational Neuroscience Annual Meeting, Los Angeles, California (2008)

Itti, L., Baldi, P.: A principled approach to detecting surprising events in video. In: Preceedings of the IEEE Conference on Computer Vision and Pattern Recognition (CVPR), vol. 1, pp. 631–637 (2005), doi:10.1109/CVPR.2005.40

Itti, L., Koch, C.: Computational modeling of visual attention. Nature Review Neuroscience 2(3), 194–203 (2001a)

Itti, L., Koch, C.: Feature combination strategies for saliency-based visual attention systems. Journal of Electronic Imaging 10(1), 161–169 (2001b)

Itti, L., Koch, C., Niebur, E.: A model of saliency-based visual attention for rapid scene analysis. IEEE Transactions on Pattern Analysis and Machine Intelligence 20(11), 1254–1259 (1998), doi:10.1109/34.730558

Itti, L., Rees, G., Tsotsos, J.: Neurobiology of Attention, 1st edn. Elsevier Press, Amsterdam (2005)

Jacob, L., Bach, F., Vert, J.P.: Clustered multi-task learning: A convex formulation. In: Advances in Neural Information Processing Systems (NIPS), pp. 745–752 (2009)

Judd, T., Ehinger, K., Durand, F., Torralba, A.: Learning to predict where humans look. In: Proceedings of the IEEE International Conference on Computer Vision (ICCV), pp. 2106–2113 (2009), doi:10.1109/ICCV.2009.5459462

Kienzle, W., Scholkopf, B., Wichmann, F.A., Franz, M.O.: How to find interesting locations in video: a spatiotemporal interest point detector learned from human eye movements. In: Proceedings of the 29th DAGM Symposium, pp. 405–414 (2007a)

Kienzle, W., Wichmann, F.A., Scholkopf, B., Franz, M.O.: A nonparametric approach to bottom-up visual saliency. In: Advances in Neural Information Processing Systems (NIPS), pp. 689–696 (2007b)

Li, J., Tian, Y., Huang, T., Gao, W.: A dataset and evaluation methodology for visual saliency in video. In: Preceedings of the IEEE International Conference on Multimedia and Expo (ICME), pp. 442–445 (2009), doi:10.1109/ICME.2009.5202529

Li, J., Tian, Y., Huang, T., Gao, W.: Probabilistic multi-task learning for visual saliency estimation in video. International Journal of Computer Vision 90(2), 150–165 (2010), doi:10.1007/s11263-010-0354-6

Li, J., Tian, Y., Huang, T.: Visual saliency with statistical priors. International Journal of Computer Vision, 1–15 (2013), doi:10.1007/s11263-013-0678-0

Liu, T., Sun, J., Zheng, N.N., Tang, X., Shum, H.Y.: Learning to detect a salient object. In: Proceedings of the IEEE Conference on Computer Vision and Pattern Recognition (CVPR), pp. 1–8 (2007), doi:10.1109/CVPR.2007.383047

Liu, T., Zheng, N., Ding, W., Yuan, Z.: Video attention: Learning to detect a salient object sequence. In: Proceedings of the 19th IEEE Conference on Pattern Recognition (ICPR), pp. 1–4 (2008), doi:10.1109/ICPR.2008.4761406

Lu, Y., Zhang, W., Jin, C., Xue, X.: Learning attention map from images. In: Preceedings of the IEEE Conference on Computer Vision and Pattern Recognition (CVPR), pp. 1067–1074 (2012), doi:10.1109/CVPR.2012.6247785

Marat, S., Ho Phuoc, T., Granjon, L., Guyader, N., Pellerin, D., Guérin-Dugué, A.: Modelling spatio-temporal saliency to predict gaze direction for short videos. International Journal of Computer Vision 82(3), 231–243 (2009), doi:10.1007/s11263-009-0215-3

Mozer, M.C., Shettel, M., Vecera, S.: Top-down control of visual attention: A rational account. In: Advances in Neural Information Processing Systems (NIPS), pp. 923–930 (2005)

Navalpakkam, V., Itti, L.: Search goal tunes visual features optimally. Neuron 53, 605–617 (2007)

Parikh, D., Zitnick, C., Chen, T.: Determining patch saliency using low-level context, Berlin, Germany, vol. 2, pp. 446–459 (2008)

Peters, R., Itti, L.: Beyond bottom-up: Incorporating task-dependent influences into a computational model of spatial attention. In: Proceedings of the IEEE Conference on Computer Vision and Pattern Recognition (CVPR), pp. 1–8 (2007), doi:10.1109/CVPR.2007.383337

Riche, N., Mancas, M., Gosselin, B., Dutoit, T.: Rare: A new bottom-up saliency model. In: Preecedings of the 19th IEEE International Conference on Image Processing (ICIP), pp. 641–644 (2012), doi:10.1109/ICIP.2012.6466941

Torralba, A., Oliva, A., Castelhano, M., Henderson, J.: Contextual guidance of eye movements and attention in real-world scenes: The role of global features on object search. Psychological Review 113(4), 766–786 (2006)

Tseng, P.H., Carmi, R., Cameron, I.G.M., Munoz, D.P., Itti, L.: Quantifying center bias of observers in free viewing of dynamic natural scenes. Journal of Vision 9(7):4, 1–16 (2009), doi:10.1167/9.7.4

Walther, D., Koch, C.: Modeling attention to salient proto-objects. Neural Networks 19(9), 1395–1407 (2006)

Wang, W., Wang, Y., Huang, Q., Gao, W.: Measuring visual saliency by site entropy rate. In: Proceedings of the IEEE Conference on Computer Vision and Pattern Recognition (CVPR), pp. 2368–2375 (2010), doi:10.1109/CVPR.2010.5539927

Wolfe, J.M.: Visual search. In: Pashler, H.E. (ed.) Attention, pp. 13–73. Psychology Press, Hove (1998)

Wolfe, J.M.: Guidance of visual search by preattentive information. In: Itti, L., Rees, G., Tsotsos, J. (eds.) Neurobiology of Attention, 1st edn., pp. 101–104. Elsevier Press, Amsterdam (2005)

Wolfe, J.M., Alvarez, G.A., Horowitz, T.S.: Attention is fast but volition is slow. Nature 406, 691 (2000)

Zhai, Y., Shah, M.: Visual attention detection in video sequences using spatiotemporal cues. In: Proceedings of the 14th Annual ACM International Conference on Multimedia, MULTIMEDIA 2006, pp. 815–824. ACM, New York (2006), doi:10.1145/1180639.1180824

Zhang, L., Tong, M.H., Marks, T.K., Shan, H., Cottrell, G.W.: Sun: A bayesian framework for saliency using natural statistics. Journal of Vision 8(7):32, 1–20 (2008), doi:10.1167/8.7.32

Zhao, Q., Koch, C.: Learning visual saliency by combining feature maps in a nonlinear manner using adaboost. Journal of Vision 12(6):22, 1–15 (2012), doi:10.1167/12.6.22

Chapter 6
Mining Cluster-Specific Knowledge for Saliency Ranking

In this Chapter, we aim to explore how to mine the prior knowledge for various scene clusters. Moreover, we also propose to model the problem of visual saliency computation in a learning to rank framework, which is proved to be more effective than the classification and regression framework.

6.1 Overview

Visual saliency plays an important role in various video applications such as video retargeting and intelligent video advertising. However, existing visual saliency computation approaches often construct a unified model for all scenes, thus leading to poor performance for the scenes with diversified contents. To solve this problem, this Chapter describes a multi-task learning to rank approach which can be used to infer multiple saliency models that apply to different scene clusters. The main contributions of this approach are summarized as follows:

1. We formulate the problem of visual saliency computation in a pair-wise learning to rank framework. In this framework, the model can automatically select the visual features that best distinguish salient targets from distractors.
2. We present an approach to construct multiple visual saliency models that apply to various scene clusters. With these scene-specific models, different features can be optimally selected and integrated to distinguish targets from distractors in different scenes.
3. We propose a multi-task learning approach to infer multiple saliency models simultaneously. By an appropriate sharing of information across models, the generalization ability of each saliency model can be greatly improved.

6.2 Multi-task Learning to Rank for Visual Saliency Computation

In this Section, we will describe the details of the proposed approach. In Sect. 6.2.1, we will clarify why the cluster-specific model is necessary for visual saliency computation. In Sect. 6.2.2, we formulate the problem into a multi-task learning to rank framework. Under this framework, a learning algorithm is proposed in Sect. 6.2.3 to simultaneously train multiple visual saliency models, each of which can adapt to a specific category for distinguishing targets and distractors. Finally, extensive experiments are conducted in Sect. 6.2.4 to demonstrate the effectiveness of the proposed approach.

6.2.1 Introduction

In natural scenes, the complexity of visual stimuli usually exceeds the processing capacity of human vision system. As a consequence, important visual subsets will be selected and processed with higher priorities. In the selective mechanism, visual saliency often plays an essential role in determining which subset (e.g., pixel, block, region or object) in a scene is important. Therefore, the central task in visual saliency computation is to rank various visual subsets in a scene to indicate their importance and processing priorities.

In visual saliency computation, each visual subset in a scene can be represented by a set of visual features. According to the Feature Integration Theory (Treisman and Gelade, 1980), different visual features can be bound into consciously experienced wholes for visual saliency computation. As such, visual saliency can be estimated by integrating the visual features that are able to effectively distinguish salient targets from distractors. Toward this end, visual saliency computation should solve two problems: what features can distinguish targets from distractors in a scene and how to optimally integrate these features.

In existing work on visual saliency computation, these two problems have been tentatively studied. Some stimuli-driven approaches (Itti et al, 1998; Itti and Koch, 2001; Itti and Baldi, 2005; Hou and Zhang, 2007; Guo et al, 2008; Harel et al, 2007; Zhai and Shah, 2006) selected the preattentive visual features and integrated them in an ad-hoc manner. In contrast, some learning-based approaches (Kienzle et al, 2007; Navalpakkam and Itti, 2007; Peters and Itti, 2007) adopted the machine learning algorithms to learn the discriminant visual features and feature integration strategies. Generally speaking, these approaches can obtain impressive results in some cases but meanwhile may suffer poor performance in other cases since they often construct a unified model for all scenes. Actually, the features that can best distinguish targets from distractors may vary remarkably in different scenes. In surveillance video, for instance, the motion features can be used to efficiently pop-out a car or a walking person (as shown in Fig. 6.1(a)-(b)); while to distinguish a red apple/flower from its surroundings, color contrasts should be used (as shown in Fig. 6.1(c)-(d)). In most cases, it is infeasible to pop-out the targets and suppress the distractors by using a fixed set of visual features. Therefore, it is necessary to construct scene-specific models that adaptively adopt different solutions for different scene categories.

    (a)   (b)   (c)   (d)

Fig. 6.1 Targets and distractors in different scenes can be best distinguished by different features. (a)-(b) the "motion" feature; (c)-(d) the "color" feature. Copyright © IEEE. All rights reserved. Reprinted, with permission, from (Li et al, 2011).

To sum up, different visual features should be adopted in different scenes to effectively pop-out targets and inhibit distractors. From the perspective of visual saliency computation, we should build scene-adaptive model to optimize the processing of various scenes. However, building the scene-specific model usually suffers from the over-fitting problem and lacks the ability to generalize. Thus a feasible solution is to build models for each scene cluster to enhance the generalization ability, especially when the training scenes are very limited.

Toward this end, we propose a multi-task learning to rank approach for visual saliency computation. Since the saliency values correspond to the processing priorities of different visual subsets in a scene, this approach formulate the problem of visual saliency computation in a learning to rank framework. Typically, a ranking model can assign an integer rank for each visual subset to indicate its processing priority, which corresponds to visual saliency to some extent. Since the best features to distinguish targets from distractors can vary remarkably in different scenes, visual saliency can be estimated by inferring multiple ranking models that apply to various scene clusters. Thus we propose a multi-task learning algorithm to infer multiple saliency models simultaneously. Different from the traditional single-task learning approach, the multi-task learning approach can carry out multiple training tasks simultaneously with fewer training data per task (Argyriou et al, 2007; Jacob et al, 2009). In this framework, the appropriate sharing of information across training tasks can be used to improve the generalization abilities and avoid overfitting. Extensive experiments on a public video benchmark reveal that this approach outperforms several state-of-the-art bottom-up, top-down and learning to rank approaches in visual saliency computation.

6.2.2 Problem Formulation

As aforementioned, the central task in visual saliency computation is to rank various visual subsets in a scene to indicate their importance and processing priorities. From the perspective of visual search, users tend to search the desired targets under the facilitation of experience derived from past similar scenes (i.e., the contextual cueing effect Chun, 2005; Torralba, 2005). Therefore, a saliency model can be represented as a ranking model which ranks all the visual subsets in a scene with respect to their relevance to the searching intention. These ranks, viewed as the priorities in searching (and processing) the desired targets, correspond to visual saliency to some extent. Without loss of generality, a subset denotes a macro-block in the remainder of this Chapter.

Given a scene S_k, we can represent its visual subsets $\{s_{kn} \in S_k\}_{n=1}^{N}$ with the local visual attributes $\{\mathbf{x}_{kn} \in \chi\}_{n=1}^{N}$. Thus the goal of visual saliency computation can be described as identifying the ranks (searching/processing properties) of $\{s_{kn}\}_{n=1}^{N}$ with respect to $\{\mathbf{x}_{kn}\}_{n=1}^{N}$. Toward this end, we infer a ranking function $\phi : \chi \rightarrow \mathbb{R}$ from past similar scenes (and user feedbacks) to assign real scores to $\{s_{kn}\}_{n=1}^{N}$. After that, these real scores can be used to rank the visual subsets. That is, $\phi(\mathbf{x}_{ku}) > \phi(\mathbf{x}_{kv})$ indicates that s_{ku} ranks higher than s_{kv} and maintains a higher saliency. Note that here the actual numerical value of $\phi(\mathbf{x})$ is immaterial and only the ordering is meaningful.

However, it is often difficult to obtain a unified ranking function that is robust to all scenes, particularly for those scenes with diversified contents. Therefore, we assume that scenes can be grouped into clusters $\{\mathbb{S}_m\}_{m=1}^{M}$ and different clusters correspond to different ranking functions $\{\phi_m\}_{m=1}^{M}$ (as shown in Fig. 6.2). Therefore, each ranking function ϕ_m should be optimized on cluster $\mathbb{S}_m$ to approximate the ground-truth (integer) ranks $\{y_{kn}\}_{n=1}^{N}$ with the estimated (integer) ranks $\{\pi_m(\mathbf{x}_{kn})\}_{n=1}^{N}$. Note that here y_{kn} can be obtained by ranking the ground-truth saliency values $\{g_{kn}\}_{n=1}^{N}$, while $\pi_m(\mathbf{x}_{kn})$ can be obtained by ranking the real scores $\{\phi_m(\mathbf{x}_{kn})\}_{n=1}^{N}$. For any two subsets, s_{ku} is more salient than s_{kv} if $g_{ku} > g_{kv}$.

Since the visual world is highly structured and such regularities can be consulted to guide visual processing (Chun, 2005), we can group K training scenes according to their global scene characteristics $\{\mathbf{v}_k\}_{k=1}^{K}$. Thus the problem of visual saliency computation can be formulated as inferring ranking functions $\Phi = \{\phi_m\}$ and scene-cluster labels $\alpha = \{\alpha_{km}\}$ from local visual attributes $\{\mathbf{x}_{kn}\}$, global scene characteristics $\{\mathbf{v}_k\}$ and ground-truth saliency values $\{g_{kn}\}$. Note that here $\alpha_{km} \in \{0, 1\}$ while $\alpha_{km} = 1$ indicates $S_k \in \mathbb{S}_m$.

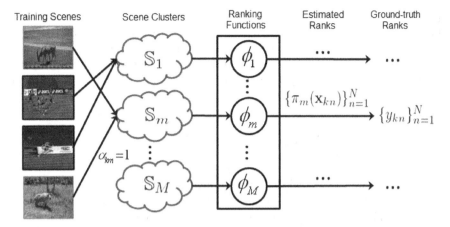

Fig. 6.2 System framework of the approach in (Li et al, 2011). Copyright © IEEE. All rights reserved. Reprinted, with permission, from (Li et al, 2011).

6.2.3   The Multi-task Learning to Rank Approach

In this Section, we will describe the details of the multi-task learning to rank approach for visual saliency computation. First, we will present how to extract the local visual attributes and global scene characteristics. After that, we propose the multi-task learning to rank approach, followed by the learning algorithm for optimizing visual saliency models and the computational complexity analysis.

6.2.3.1   Feature Extraction

First, we will introduce how to calculate the local visual attributes $\mathbf{x}_{kn}$ and the global scene characteristics $\mathbf{v}_k$. Often, the contrast-based visual irregularities in preattentive features can well recover the salient locations in a scene. By using the algorithm proposed in (Itti and Koch, 2001), we compute "center-surround" local contrasts for each visual subset in a scene. Typically, such local contrasts are robust to noise and quality degradation such as brightness changing, quality compression and blurring. In the computation, the local contrasts are generated from 12 preattentive visual channels in 6 scales, including intensity (6), red/green and blue/yellow opponencies (12), four orientations (24), temporal flickers (6) and motion energies in four directions (24). In total $L=72$ local contrast features are obtained to form the local visual attributes $\mathbf{x}_{kn}$.

As proposed in (Chun, 2005), some stable properties of the visual environment, such as rough spatial layouts and predicable variations, do not change radically over time and can be encoded to guide the visual processing. These properties can work as the contextual priors for individuals to search and process the targets in similar environments. Thus we use them to form a global descriptor to characterize a scene, which can be obtained by summarizing the local visual attributes of a scene without encoding specific objects or regions (Torralba, 2005). Here we adopt the mean and standard deviation of the lth dimension of $\mathbf{x}_{kn}$, which are computed as:

$$\mu_{kl} = \frac{1}{N} \sum_{n=1}^{N} \mathbf{x}_{kn}\left(l\right), \sigma_{kl} = \sqrt{\frac{1}{N} \sum_{n=1}^{N} \left(\mathbf{x}_{kn}\left(l\right) - \mu_{kl}\right)^2}. \qquad (6.1)$$

The mean and standard deviation indicate whether visual subsets in the scene S_k are discriminative from each other in the lth feature. After the calculation, $\{\mu_{kl}\}_{l=1}^{L}$ and $\{\sigma_{kl}\}_{l=1}^{L}$ can be used to form the global scene descriptor $\mathbf{v}_k$ with $2L$ components. Since different visual features may span different ranges of values, we normalize each dimension of $\mathbf{x}_{kn}$ and $\mathbf{v}_k$ into $[0, 1]$.

6.2.3.2   Multi-task Learning to Rank Approach

Given the local and global features, we can group K training scenes into M clusters and infer M ranking functions in a multi-task learning framework. Without loss of generality, we define $\phi_m(\mathbf{x}) = \omega_m^{\mathbf{T}}\mathbf{x}$ since various preattentive features are often integrated into experienced wholes with linear weights for saliency computation (e.g., Itti et al, 1998; Itti and Koch, 2001; Itti and Baldi, 2005; Navalpakkam and Itti, 2007). Note that ω_m is a column vector with L components. For the sake of simplicity, we let $\mathbf{W}$ be a $L \times M$ matrix with the mth column equals to ω_m. Therefore, the optimization objective can be defined as:

$$\min_{\mathbf{W},\alpha} \mathcal{L}\left(\mathbf{W}, \alpha\right) + \Omega\left(\mathbf{W}, \alpha\right),$$

$$s.t. \sum_{m=1}^{M} \alpha_{km} = 1, \forall k \text{ and } \alpha_{km} \in \{0, 1\}, \forall m, \qquad (6.2)$$

where $\mathcal{L}\left(\mathbf{W}, \alpha\right)$ is the empirical loss and $\Omega\left(\mathbf{W}, \alpha\right)$ is the penalty term that encodes the prior knowledge on the parameters. To focus on the features that can distinguish targets from distractors, we define the empirical loss $\mathcal{L}\left(\mathbf{W}, \alpha\right)$ in a pair-wise manner:

$$\mathcal{L}\left(\mathbf{W}, \alpha\right) = \sum_{k=1}^{K} \sum_{m=1}^{M} \alpha_{km} \sum_{u \neq v}^{N} [g_{ku} < g_{kv}]_{\mathbf{I}} [\omega_m^{\mathbf{T}}\mathbf{x}_{ku} \geq \omega_m^{\mathbf{T}}\mathbf{x}_{kv}]_{\mathbf{I}}, \qquad (6.3)$$

where $[x]_{\mathbf{I}}=1$ if x holds, otherwise $[x]_{\mathbf{I}}=0$. We can see that the empirical loss equals to the number of falsely ranked subset pairs on all training scenes.

Beyond the empirical loss, the prior knowledge on grouping scenes and training ranking functions should also be considered in the optimization process. That is, the optimization objective should comprise the penalty terms that encode the priors on scene clustering, model correlation and model complexity. These three penalty terms can be defined as:

1. **Scene clustering**. To group scenes with similar contents into the same cluster, we set the penalty term Ω_s as:

$$\Omega_s = \frac{1}{K} \sum_{k_0 \neq k_1}^{K} \sum_{m=1}^{M} (\alpha_{k_0 m} - \alpha_{k_1 m})^2 \cos(\mathbf{v}_{k_0}, \mathbf{v}_{k_1}), \qquad (6.4)$$

where $\cos(\mathbf{v}_{k_0}, \mathbf{v}_{k_1})$ denotes the similarity of the k_0th and the k_1th scenes, which is computed as the cosine distance between two vectors $0 \preceq \mathbf{v}_{k_0}, \mathbf{v}_{k_1} \preceq 1$. We can see that the penalty term will be large when two similar scenes S_{k_0} and S_{k_1} are grouped into different clusters.

2. **Model correlation**. In the training process, each scene cluster may only contain limited number of scenes with similar contents. Thus the saliency models directly trained on these clusters may lack the generalization ability (as shown in Fig. 6.3(a), this corresponds to the typical single task learning). To solve this problem, taking an appropriate sharing of information across training tasks can avoid overfitting and improve the performance of each model (Jacob et al, 2009; Evgeniou et al, 2005). Therefore, we set a penalty term Ω_d to incorporate the correlations between models:

$$\Omega_d = \frac{1}{M} \sum_{i \neq j}^{M} \sum_{k=1}^{K} \sum_{u \neq v}^{N} [g_{ku} < g_{kv}]_{\mathbf{I}} \times$$
$$[\omega_i^{\mathbf{T}} \mathbf{x}_{ku} \geq \omega_i^{\mathbf{T}} \mathbf{x}_{kv}]_{\mathbf{I}} [\omega_j^{\mathbf{T}} \mathbf{x}_{ku} \geq \omega_j^{\mathbf{T}} \mathbf{x}_{kv}]_{\mathbf{I}}. \qquad (6.5)$$

The influence of this penalty is two-fold. First, a sample mistakenly predicted by most models will be emphasized in training the model ϕ_i (i.e., a large $\sum_{j \neq i} [g_{ku} < g_{kv}]_{\mathbf{I}} [\omega_j^{\mathbf{T}} \mathbf{x}_{ku} \geq \omega_j^{\mathbf{T}} \mathbf{x}_{kv}]_{\mathbf{I}}$). This ensures the diversity of training samples for ϕ_i, leading to improved generalization ability. Second, a sample successfully predicted by most models will be ignored in training ϕ_i (i.e., a small $\sum_{j \neq i} [g_{ku} < g_{kv}]_{\mathbf{I}} [\omega_j^{\mathbf{T}} \mathbf{x}_{ku} \geq \omega_j^{\mathbf{T}} \mathbf{x}_{kv}]_{\mathbf{I}}$). This guarantees the diversity of different models. With this penalty term, each model is actually related to all the training samples with different weights (as shown in Fig. 6.3(b)), leading to improved performance.

3. **Model complexity**. To avoid optimizing complex models, we have to set a penalty term Ω_c as:

$$\Omega_c = \sum_{m=1}^{M} \omega_m^{\mathbf{T}} \omega_m. \qquad (6.6)$$

Here the penalty term on model complexity can be used to constrain the number of scene clusters. Thus over-complex models can be avoided.

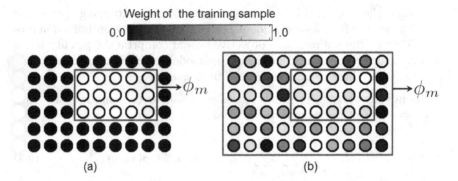

Fig. 6.3 The advantage of the multi-task learning approach. (a) In single task learning, each model is trained independently on limited samples (i.e., those in the red box) and may lack the generalization ability; (b) In multi-task learning, each model is actually trained on the whole dataset (i.e., the samples in the blue box) by emphasizing different subset of samples (i.e., the samples in the red box), leading to improved generalization ability. Copyright © IEEE. All rights reserved. Reprinted, with permission, from (Li et al, 2011).

With these penalty terms, the overall penalty $\Omega\left(\mathbf{W},\alpha\right)$ can be written as the weighted linear combination of them:

$$\Omega(\mathbf{W},\alpha) = \epsilon_s\Omega_s + \epsilon_d\Omega_d + \epsilon_c\Omega_c, \qquad (6.7)$$

where ϵ_s, ϵ_d and ϵ_c are three non-negative weights to combine these penalty terms. In this study, these weights are empirically selected using the validation set.

6.2.3.3   The Learning Algorithm

By incorporating the empirical loss in (6.3) and penalty term in (6.7) into the optimization objective (6.2), we will encounter a non-convex optimization problem. Therefore, we use the EM algorithm (Dempster et al, 1977) to iteratively solve the problem and ensure the convergence. First, scenes are simply grouped into M clusters according to their global scene characteristics using the K-means algorithm (M can be empirically selected through cross validation). After that, the scene-cluster labels α are initialized and the parameter matrix $\mathbf{W}$ can be initialized by minimizing (6.2). Here we set $\epsilon_s = \epsilon_d = \epsilon_c = 0$ to only minimize the empirical loss without considering the inter-scene and inter-model correlations (this is also the baseline used in the experiments). After the initialization, we can iteratively update α and optimize $\mathbf{W}$ using the EM-algorithm. As in (Freund et al, 2003; Boyd and Vandenberghe, 2008), we replace the Boolean terms related to $\mathbf{W}$ in (6.2) with their convex upper bounds to facilitate the optimization:

$$[\omega_m^{\mathbf{T}}\mathbf{x}_{ku} \geq \omega_m^{\mathbf{T}}\mathbf{x}_{kv}]_{\mathbf{I}} \leq \exp\left(\omega_m^{\mathbf{T}}\mathbf{x}_{ku} - \omega_m^{\mathbf{T}}\mathbf{x}_{kv}\right)$$
$$= \exp\left(trace\left(\mathbf{W}^{\mathbf{T}}\mathbf{X}_{kuv}^m\right)\right), \tag{6.8}$$

where $\mathbf{X}_{kuv}^m$ is a $L \times M$ matrix with its mth column equals to $\mathbf{x}_{ku} - \mathbf{x}_{kv}$ and the other components equal to zero. Here we adopt the exponential upper bound since it is convex (although loose) and can facilitate the optimization. For the sake of simplicity, we let:

$$\eta_{kuv}^m = \exp\left(trace\left(\mathbf{W}^{\mathbf{T}}\mathbf{X}_{kuv}^m\right)\right). \tag{6.9}$$

After the replacement, we can iteratively update α and optimize $\mathbf{W}$ as:
Step 1: For $k = 1, \ldots, K$, update $\alpha_k = [\alpha_{k1}, \ldots, \alpha_{kM}]$ by solving the problem which contains only the terms in (6.2) that are related to α_k:

$$\min_{\alpha_k} \sum_{m=1}^{M} \alpha_{km} \sum_{u \neq v}^{N} [g_{ku} < g_{kv}]_{\mathbf{I}} \eta_{kuv}^m$$

$$+ \frac{2\epsilon_s}{K} \sum_{k_0 \neq k}^{K} \sum_{m=1}^{M} (\alpha_{km} - \alpha_{k_0 m})^2 \cos(\mathbf{v}_k, \mathbf{v}_{k_0}), \tag{6.10}$$

$$s.t. \sum_{m=1}^{M} \alpha_{km} = 1 \text{ and } \alpha_{km} \in \{0, 1\}, \forall m.$$

The optimization objective contains only quadratic terms with linear constraints and can be efficiently solved by 0-1 programming. In the optimization, the first term indicates that a scene will be grouped into the cluster with low prediction error. The second term indicates that such process is also influenced by the labels of similar scenes.

Step 2: To optimize $\mathbf{W}$, we have to solve the problem which contains only the terms in (6.2) that are related to $\mathbf{W}$:

$$\min_{\mathbf{W}} \sum_{k=1}^{K} \sum_{u \neq v}^{N} [g_{ku} < g_{kv}]_{\mathbf{I}} \sum_{m=1}^{M} \alpha_{km} \eta_{kuv}^m$$

$$+ \frac{\epsilon_d}{M} \sum_{k=1}^{K} \sum_{u \neq v}^{N} [g_{ku} < g_{kv}]_{\mathbf{I}} \sum_{i \neq j}^{M} \eta_{kuv}^i \eta_{kuv}^j \tag{6.11}$$

$$+ \epsilon_c trace\left(\mathbf{W}^{\mathbf{T}}\mathbf{W}\right).$$

Since the exponential upper bound is convex, the objective function (6.11) turns to be convex since it contains only quadratic and exponential terms of

W with non-negative weights. Therefore, we can solve it with gradient descent method. Note that we have:

$$\frac{\partial \eta_{kuv}^m}{\partial \mathbf{W}} = \frac{\partial \exp\left(trace\left(\mathbf{W}^{\mathbf{T}} \mathbf{X}_{kuv}^m\right)\right)}{\partial \mathbf{W}} = \mathbf{X}_{kuv}^m \eta_{kuv}^m. \tag{6.12}$$

Therefore, the gradient direction can be written as:

$$\begin{aligned}
\triangle \mathbf{W} \propto 2\epsilon_c \mathbf{W} + \sum_{k=1}^{K} \sum_{u \neq v}^{N} [g_{ku} < g_{kv}]_{\mathbf{I}} \sum_{m=1}^{M} \mathbf{X}_{kuv}^m \\
\times \left(\alpha_{km} \eta_{kuv}^m + \frac{2\epsilon_d}{M} \sum_{i \neq m}^{M} \eta_{kuv}^i \eta_{kuv}^m \right).
\end{aligned} \tag{6.13}$$

From (6.13), we can see that each model is actually optimized on the whole training set by emphasizing different subset of training samples (as shown in Fig. 6.3(b)). Actually, the optimization process of **W** mainly involves two steps. That is, estimating the prediction of each ranking function on each training sample to update η_{kuv}^m and re-weighting each training sample to calculate the gradient direction $\triangle\mathbf{W}$. By iteratively performing these two steps, the convex objective (6.11) can effectively reach its global minimum.

The detailed learning process is listed in Algorithm 1. By iteratively updating α and optimizing **W**, we can obtain a decreasing overall loss. In the algorithm, we iteratively carry out these two steps until the algorithm converges or reaches a predefined number of iterations.

Algorithm 1. The multi-task learning to rank algorithm

Input : Local visual attributes $\{\mathbf{x}_{kn}\}$, global scene characteristics $\{\mathbf{v}_k\}$,
        ground-truth saliency $\{g_{kn}\}$, iteration times T, threshold $\triangle e$.
Output: Model parameters **W**, scene-cluster labels α.
begin
    Initialization:
    Group scenes into M clusters using K-means;
    Initialize α;
    Initialize **W** by minimizing (6.11) (set $\epsilon_s = \epsilon_d = \epsilon_c = 0$);
    $e^{(1)} \leftarrow \mathcal{L}(\mathbf{W}, \alpha) + \Omega(\mathbf{W}, \alpha)$;
    $t \leftarrow 1$;
    Optimization:
    repeat
        for $k \leftarrow 1$ **to** K **do** $\alpha_k^* \leftarrow$ minimize (6.10);
        $\mathbf{W} \leftarrow$ minimize (6.11);
        $e^{(t)} \leftarrow \mathcal{L}(\mathbf{W}, \alpha) + \Omega(\mathbf{W}, \alpha)$;
        $t \leftarrow t + 1$;
    until $t \geq T$ *or* $(e^{(t)} - e^{(t-1)}) < \triangle e$;
end

Given a new scene, we have to identify a proper ranking function for estimating its saliency. Recall that the global descriptor can characterize a scene and guide the visual processes on it (Chun, 2005), we can select the ranking function for the new scene based on its global scene characteristics. That is, scenes with similar global characteristics are supposed to undergo the similar visual processes and the same ranking functions can be used for the saliency computation. Therefore, we adopt a KNN classifier (1-NN in this study) to find the scene in the training set with the most similar global scene characteristics (i.e., the cosine similarity as in (6.4)). After that, the corresponding ranking function can be selected for visual saliency computation.

In order to test the performance of the learned visual saliency model, we have to compare the estimated ranks with eye fixation data. However, it is often difficult to directly perform such comparison. Therefore, we generate the estimated saliency map from these ranks and the ground-truth saliency map from eye fixations for further comparison (as shown in Fig. 7.6). Toward this end, we let $c_{kn} \in \{1, \ldots, N\}$ be the rank for the visual subset s_{kn} and empirically transform c_{kn} to the estimated saliency value $\widehat{e}_{kn}$:

$$\widehat{e}_{kn} = G(0, \sigma) * \left(\frac{N - c_{kn}}{N} \right)^{\beta}, \tag{6.14}$$

where $\beta > 0$ is a constant to pop-out the most salient targets and can be selected using the validation set. For larger β, the distractors can be suppressed more effectively. In the experiments, we set $\beta = 5$ and only limited locations can pop-out. In (6.14), a Gaussian kernel $G(0, \sigma)$ is adopted to generate the estimated saliency map from these locations by modeling the decrease in accuracy of the fovea with increasing eccentricity. Here we set $\sigma = 5$. For fair comparison, the same kernel is also used in the experiments to construct the ground-truth saliency maps from eye fixation data. For convenience, the estimated saliency values for all visual subsets in a scene are normalized into $[0, 1]$.

6.2.3.4 Computational Complexity Analysis

In terms of computational complexity, the cost of the EM optimization, C_{EM}, can be written as:

$$C_{EM} = \sum_{i=1} C_\alpha^i + C_{\mathbf{W}}^i. \tag{6.15}$$

where C_α^i ($C_{\mathbf{W}}^i$) denote the cost of updating α (optimizing $\mathbf{W}$) in the ith iteration. Thus we can calculate these two costs separately to obtain the overall cost. For the sake of convenience, we let $N_a^k = \sum_{u \neq v}^N [g_{ku} < g_{kv}]_{\mathbf{I}}$ be the number of training samples in the kth scene and $N_a = \sum_{k=1}^K N_a^k$ be the total number of samples in all K scenes.

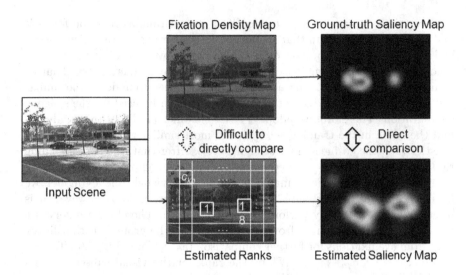

Fig. 6.4 Integer ranks are turned into real values for direct comparison. Copyright © IEEE. All rights reserved. Reprinted, with permission, from (Li et al, 2011).

In updating α, the cosine distance of any two scenes can be obtained before the EM optimization and we suppose that $\{\eta_{kuv}^m\}$ have already been calculated (e.g., in optimizing $\mathbf{W}$). Therefore, α_k can be updated by taking the influences of the other K-1 scenes into account. With respect to (6.10), the complexity of updating α_k is $O(N_a^k M + KM)$. Therefore, the overall complexity of updating $\alpha = \{\alpha_k\}$ is:

$$C_\alpha^i = \sum_{k=1}^{K} O(N_a^k M + KM) = O(N_a M + K^2 M). \qquad (6.16)$$

When optimizing $\mathbf{W}$, the cost depends on the convergence rate of the gradient-descent algorithm as well as the complexity for each gradient step. Let R_i be the number of gradient steps in optimizing $\mathbf{W}$ in the ith EM iteration and $C_\triangle^i$ be the computational complexity in each gradient step. Typically, R_i may vary remarkably in each EM iteration while the complexity $C_\triangle^i$ often remains constant in each gradient step. Therefore, the computational cost $C_\mathbf{W}^i$ can be written as:

$$C_\mathbf{W}^i = R_i \times C_\triangle^i. \qquad (6.17)$$

From (6.13), we observe that $C_\triangle^i$ is related to two complexities:
▶ The complexity in computing $\triangle \mathbf{W}$ using (6.13). Recall that only the mth column in $\mathbf{X}_{kuv}^m$ may have non-zero components (i.e., this column equals to $\mathbf{x}_{ku}$-$\mathbf{x}_{kv}$), the multiplication of $\mathbf{X}_{kuv}^m$ with a real value and the addition of

$\mathbf{X}_{kuv}^m$ with another matrix have $O(L)$ complexity. Therefore, computing $\triangle \mathbf{W}$ using (6.13) has approximately $O(N_a M^2 L)$ complexity.

▶ The complexity in computing $\{\eta_{kuv}^m\}$ using (6.9). From (7.5) and (6.9), we observe that $\eta_{kuv}^m = \exp(\omega_m^\mathbf{T} \mathbf{x}_{ku} - \omega_m^\mathbf{T} \mathbf{x}_{kv})$. Thus the complexity of updating $\{\eta_{kuv}^m\}$ is $O(N_a M L)$.

With these two complexities, $C_\triangle^i$ can be written as:

$$C_\triangle^i = O(N_a M^2 L) + O(N_a M L) \approx O(N_a M^2 L). \qquad (6.18)$$

By incorporating (6.16)-(6.18) into (6.15), we observe that the overall computational complexity is tightly correlated with six parameters, including: K (the number of training scenes), N_a (the number of training samples), M (the number of scene clusters), L (local feature dimensionality), $\{R_i\}$ (the numbers of gradient steps in optimizing $\mathbf{W}$) and the number of EM iterations. Among these six parameters, K is determined by the training set and different gradient-descent algorithms have different convergence rates, leading to different $\{R_i\}$ (Watrous, 1987). Moreover, we observe in the experiment that the EM optimization usually terminates in less than $T=10$ iterations. For the other three parameters, there are three feasible ways to reduce the computational complexity:

1. Remove the redundant training samples to reduce N_a;
2. Reduce the cluster number M and
3. Reduce the features dimensionality L.

Often, the parameter L is predefined in different application (i.e., the number of candidate visual features in a specific application) and M should be optimized through cross validation. Therefore, we can reduce the computational complexity by removing the redundant training samples (e.g., by fusing the subsets in each scene with similar local visual attributes and ground-truth saliency values). In the experiment, we observe that when the scene number K, feature dimensionality L and clusters number M are considered to be constants, the training time is linear with respect to the number of training samples. Compared with the typical multi-task learning approaches whose complexity may scale as the cube of the number of training samples (e.g., when using the regularization networks in Evgeniou et al, 2005; Pillonetto et al, 2009), the computational complexity of the proposed approach is much less and is thus acceptable.

6.2.4   Experiments

In this Section, we evaluate the proposed approach on a public eye-fixation dataset (Itti, 2008). The dataset, denoted as **ORIG**, consists of 46,489 frames in 50 video clips (25 minutes, 640×480). These video clips mainly contains genres such as "outdoors," "TV news," "sports," "commercials," "video games" and "talk shows." For these clips, the dataset also provides the eye traces of eight subjects recorded using a 240HZ ISCAN RK-464 eye-tracker (four to six subjects per clip). Based on these eye traces, the fixation density in each 16×16 macro-block is calculated. After that, the ground-truth saliency map for each scene can be constructed by convolving the fixation density map with a Gaussian kernel ($\sigma=5$) to model the decrease in accuracy of the fovea with increasing eccentricity.

On this dataset, the main objective of the experiment is to evaluate whether this approach can learn the features to distinguish targets from distractors and whether the learned features can be effectively transferred to new scenes for visual saliency computation. Toward this end, we randomly select 1/10 scenes from the **ORIG** dataset to construct the training set, 1/10 scenes for validation and the rest scenes are used for testing. For the sake of convenience, the training/validation/testing sets are denoted as $\mathbb{D}_{train}$, $\mathbb{D}_{validate}$ and $\mathbb{D}_{test}$, respectively. On these training/validation/testing sets, four experiments are conducted. In the first experiment, the performance of the proposed approach when using different parameters is tested. A set of optimal parameters are also selected for the proposed approach, which will be used in the other three experiments. In the second and the third experiments, this approach is compared with the state-of-the-art saliency models and ranking models, respectively. These two experiments are designed to demonstrate the performance of this approach from different perspectives. Finally, we test the performance of all these models on different scene genres of the **ORIG** dataset in the last experiment.

In these experiments, the proposed multi-task learning to rank approach (**MTLR**) is compared with 12 state-of-the-art saliency/ranking models as well as the baseline of our approach. In general, these models can be grouped into three categories:

▶ **Bottom-up models for saliency computation**:
- Itti98 (Itti et al, 1998) and Itti01 (Itti and Koch, 2001): models based on local contrasts;
- Itti05 (Itti and Baldi, 2005) and Zhai06 (Zhai and Shah, 2006): models that mainly focus on inter-frame variation;
- Harel07 (Harel et al, 2007): a graph-based model by detecting spatiotemporal irregularities;
- Hou07 (Hou and Zhang, 2007) and Guo08 (Guo et al, 2008): models based on spectral analysis;

▶ **Top-down models for saliency computation**:
- Kienzle07 (Kienzle et al, 2007): a model that learns the correlations between local features and visual saliency by using a SVM classifier;
- Navalpakkam07 (Navalpakkam and Itti, 2007): a model that combines local contrasts in preattentive features by optimizing the signal-noise-ratio.

- Peter07 (Peters and Itti, 2007): a model that learns the projection matrix between global scene characteristics and eye density maps;
▶ **Ranking models for saliency computation**:
- Freund03 (Freund et al, 2003): a boosting algorithm for learning weak rankers as well as their combination strategies;
- Joachims06 (Joachims, 2006): a pair-wise learning to rank algorithm using Support Vector Machine.
- **Base-I**: the baseline of our approach which groups scenes into clusters and infers a model for each cluster without considering the inter-scene and inter-model correlations.

Note that the ranking-based approaches such as Freund03, Joachims06 and **Base-I** can only give integer ranks, we also turn them into real saliency values using the same way as **MTLR** for fair comparisons.

In the comparisons, the receiver operating characteristics (**ROC**) curve is used for performance evaluation. In the evaluation, a set of thresholds $T_{roc} = \{0.00, 0.01, \ldots, 1.00\}$ are used to select the salient visual subsets from all the estimated saliency maps predicted by a specific saliency model. These salient subsets are then validated according to the ground-truth saliency maps. For the threshold T_{roc}, the true positives (TP), false negatives (FN), false positives (FP) and true negatives (TN) on all the *test data* can be calculated as:

$$TP = \sum_k \sum_{n=1}^{N} [\widehat{e}_{kn} \geq T_{roc}]_{\mathbf{I}} \cdot g_{kn},$$

$$FN = \sum_k \sum_{n=1}^{N} [\widehat{e}_{kn} < T_{roc}]_{\mathbf{I}} \cdot g_{kn},$$

$$FP = \sum_k \sum_{n=1}^{N} [\widehat{e}_{kn} \geq T_{roc}]_{\mathbf{I}} \cdot [g_{kn} = 0]_{\mathbf{I}},$$

$$TN = \sum_k \sum_{n=1}^{N} [\widehat{e}_{kn} < T_{roc}]_{\mathbf{I}} \cdot [g_{kn} = 0]_{\mathbf{I}},$$

(6.19)

where $0 \leq \widehat{e}_{kn}, g_{kn} \leq 1$ are the estimated and ground-truth saliency values, respectively. After that, the false positive rate is calculated as $FP/(FP+TN)$ and the true positive rate is calculated as $TP/(TP+FN)$. Correspondingly, the **ROC** curve for the saliency model is plotted as the *false positive rate* vs. *true positive rate*. Moreover, the Area Under the **ROC** Curve (**AUC**) is also calculated to demonstrate the overall performance of a saliency model. We also compute the improvement of **MTLR** against other approaches on the **AUC** score, denoted as **IMP**.

6.2.4.1   Parameter Selection

This experiment is designed to evaluate the influence of different parameters to our approach. In this experiment, we demonstrate the performance of our approach when using various parameters and select a set of optimal parameters for the next three experiments. The computational complexities when using different parameters are also reported.

In our approach, there are many parameters involved in the processes of model optimization and saliency computation. Among these parameters, K (the number of training scenes) and N (the number of visual subsets in each scene) are determined by the training set. Note that here a visual subset corresponds to a 16×16 macro-block, which is the same as the block used in calculating the ground-truth saliency maps. Other parameters, such as $\epsilon_s, \epsilon_d, \epsilon_c, M$ and β, have to be optimized by cross validation. By using the training set $\mathbb{D}_{train}$ and validation set $\mathbb{D}_{validate}$, we test the influence of each parameter by fixing all the other parameters. The influence of various parameters are summarized as follows:

▶ ϵ_s. In the EM optimization, some scene-cluster labels may vary radically with respect to the prediction errors when ϵ_s is too small. In contrast, the scene-cluster labels will rarely change when ϵ_s is too large.

▶ ϵ_d. When ϵ_d is small, **MTLR** is slightly influenced by inter-model correlations and its performance will approximate that of **Base-I**. For large ϵ_d, the models trained on various scene clusters may lack the diversity, leading to decreased performance.

▶ ϵ_c. This parameter is only used to avoid constructing over-complex models. A smaller ϵ_c indicates that a more complex model is acceptable.

▶ M. The scene cluster number M is an important parameter in the optimization process. Therefore, we draw a curve to demonstrate the influence of M to the **AUC** score. As shown in Fig. 6.5, a larger M may lead to higher **AUC** score. However, the computational complexity will become extremely high when M is become too large. Therefore, the selection of M should simultaneously consider the algorithm performance and computational complexity. In our experiment, we start from $M = 1$ and select the optimal M as the smallest number of clusters on which the **AUC** score is larger than the scores on $M + 1$, $M + 2$ and $M + 3$. Here we test the **AUC** scores on four successive cluster numbers to avoid sudden fluctuations.

Moreover, we also illustrate some typical scene clusters in Fig. 6.5. From these scene clusters, we can see that scenes with similar spatial layouts and predictable variations will be grouped together as M gets large (e.g., M=15). However, an increasing M may not always guarantee an increasing **AUC** score (e.g., the **AUC** scores in Fig. 6.5 when $M = 22$ and $M = 15$). Actually, **Base-I** may become over-fitting when M is too large, while **MTLR** can avoid such problem by utilizing the inter-scene and inter-model correlations.

▶ β. β is an important parameter to turn the estimated integer ranks into real saliency values. Therefore, we also draw a curve to demonstrate the

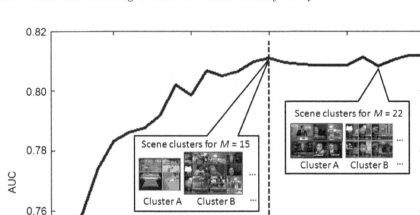

influence of β to the **AUC** score. As shown in Fig. 6.6, the **AUC** reaches its maximum around $\beta=5$. For smaller β, distractors cannot be adequately suppressed, leading to noisy saliency maps. In contrast, a large β may also suppress some salient targets.

By iteratively carrying out the cross validation and varying these parameters, a set of optimal parameters are selected for **MTLR**. Here we have $\epsilon_s=0.10$, $\epsilon_d=0.11$, $\epsilon_c=0.14$, $M=15$ and $\beta=5$. Note that the same cluster number M is also used for **Base-I** in the following experiments. Moreover, the parameters of all the other learning-based approaches, including Kienzle07, Navalpakkam07, Peters07, Freund03 and Joachims06, have their parameters optimized in the same way to get fair comparisons in the following experiments.

In addition, we also conduct an experiment to test the computational complexity of **MTLR**. With the optimal parameters listed above, we use different criteria to remove the redundant training samples (e.g., fusing the visual subsets in a scene with similar local visual attributes and ground-truth saliency values) and test the performance of **MTLR**. Note that here all the training processes are carried on a DELL Optiplex 960 computer with three threads.

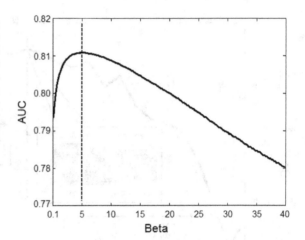

As shown in Fig. 6.7, the time costs before the convergence of the EM optimization are almost linearly correlated with the numbers of training samples. In contrast, the **AUC** scores will increase when **MTLR** is trained with more samples. However, we can see from Fig. 6.7 that reducing the training number to 30%-40% will not severely decrease the **AUC** score. Therefore, reducing N_a is a feasible way to reduce the computational complexity while preserving the overall performance.

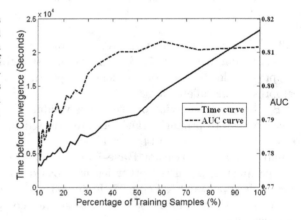

Moreover, we also test the efficiency of **MTLR** in estimating visual saliency. On a 2.66GHz CPU, the time used in each major step (i.e., extracting local/global features, selecting ranking function through 1-NN, predicting ranks and transforming these ranks into real saliency values) is recorded. Note that here the I/O time is not taken into account. On average, it takes about 0.109s to estimate the visual saliency for a new scene. From this result, we can see that our approach demonstrates high efficiency in visual saliency computation, which is very important since the saliency computation is usually the first step for many applications.

6.2.4.2   Comparisons with Saliency Models

In this experiment, we will test the performance of various visual saliency models and give explanations from the neurobiological perspective. The comparisons are mainly made between **MTLR** and ten bottom-up and top-down visual saliency models. In the experiment, the comparisons are conducted for ten times. In each comparison, the learning-based approaches are trained on $\mathbb{D}_{train}$ and tested on $\mathbb{D}_{test}$. After that, a **ROC** curve is generated for each model based on all the estimated saliency maps in all the ten comparisons. A unified **AUC** score is also reported to demonstrate the performance of each model. The **AUC** scores of various saliency models are shown in Table 6.1[1]. The **ROC** curves are illustrated in Fig. 6.8 and some representative results are given in Fig. 6.9.

Table 6.1 Performance of various saliency models on **ORIG**. Copyright © IEEE. All rights reserved. Reprinted, with permission, from (Li et al, 2011).

Algorithm		**AUC**	IMP (%)
Bottom-Up	Itti98	0.557	45.5
	Itti01	0.554	46.4
	Itti05	0.622	30.4
	Zhai06	0.637	27.3
	Harel07	0.584	38.9
	Hou07	0.666	21.7
	Guo08	0.674	20.3
Top-Down	Kienzle07	0.539	50.3
	Navalpakkam07	0.697	16.3
	Peters07	0.693	17.1
	MTLR	**0.811**	

From Table 6.1 and Fig. 6.8, we can see that **MTLR** outperforms all the other visual saliency models. As shown in Fig. 6.9(c)-(d), Itti98 and Itti01 only maintain the most salient subsets with the "winner-take-all" competition, leading to the low **AUC** scores. In contrast, Itti05, Zhai06 and Harel07 have achieved a bit improvement by focusing on the spatiotemporal visual irregularities, while Hou07 and Guo08 perform even better by detecting such irregularities through spectrum analysis. As shown in Fig. 6.9(e)-(i), these five bottom-up approaches can well locate the salient subsets but may have

---

[1] The **AUC** scores reported in Tables 5.2, 6.1 and 7.1 may change slightly due to different experimental settings (e.g., sizes of Gaussian kernels used in convolution, resolutions of saliency maps and settings of training/validation/testing sets).

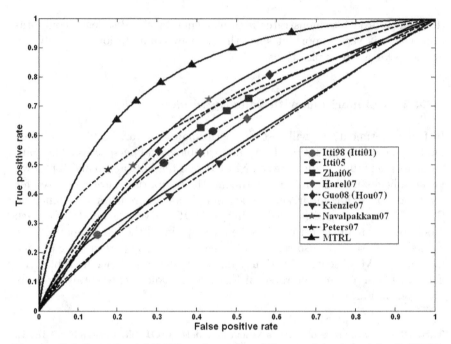

Fig. 6.8 ROC curves of **MTLR** and various saliency models. Copyright © IEEE. All rights reserved. Reprinted, with permission, from (Li et al, 2011).

difficulties in suppressing the distractors. In particular, we observe from Table 6.1 that the spatial saliency models (e.g., Itti98 and Hou07) may perform even better than some spatiotemporal saliency models (e.g., Itti01, Zhai06 and Itti05). This indicates that simply incorporating the temporal features (e.g., motion and flicker) may not always guarantee a better performance. Actually, human fixations can be driven by different spatial/temporal visual features in different scenes. Thus selecting the right features to distinguish targets from distractors is the most important issue for visual saliency computation other than incorporating more candidate visual features into the model.

Often, it is believed that the experience from past similar scenes can assist the suppression of distractors. However, the top-down approach Kienzle07 acts even worse than the bottom-up approaches, since it simply maps the local visual features to saliency values. However, the correlations between visual features and saliency values may not always hold in different scenes. In contrast, Peters07 simply infers the relations between global scene characteristics and eye density maps. As shown in Fig. 6.9(l) and Table 6.1, the advantage of Peters07 is that it can well recover the most probable salient locations (e.g., the center of each scene), leading to a higher **AUC** score. However, they may also introduce a lot of noise into the estimated saliency maps. Particularly, Peters07 have to infer an 832×300 projection matrix

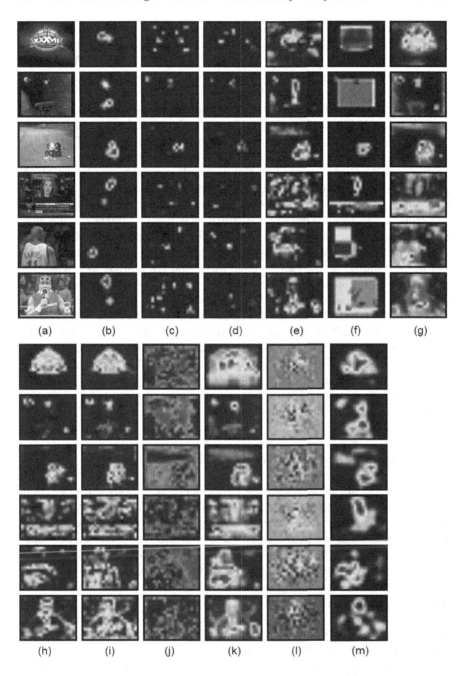

Fig. 6.9 Representative results of visual saliency models. (a) Original scenes; (b) Ground-truth saliency maps; (c) Itti98; (d) Itti01; (e) Itti05; (f) Zhai06; (g) Harel07; (h) Hou07; (i) Guo08; (j) Kienzle07; (k) Navalpakkam07; (l) Peters07; (m) **MTLR**. Copyright © IEEE. All rights reserved. Reprinted, with permission, from (Li et al, 2011).

between global scene characteristics and eye density maps, while the training data is often insufficient to do so. In fact, the **AUC** score of Peters07 can reach 0.745 if it is trained on 9/10 scenes of the **ORIG** dataset. This hampers the utilization of Peters07 in some applications which may only provide sparse training data and user feedbacks.

As shown in Table 6.1 and Fig. 6.8, we can see that Navalpakkam07 acts much better (**AUC**=0.697) by considering both the influences of local visual attributes and target-distractor correlations. Moreover, **MTLR** have achieved the best performance (**AUC**=0.811) by simultaneously taking the influences of local visual attributes and global scene characteristics into a pair-wise ranking framework. From the neurobiological perspective, the pair-wise ranking framework can assist finding features that can distinguish targets from distractors. With these features, the salient targets can be successfully pop-out while distractors can be effectively suppressed. Meanwhile, the experience of successfully popping-out targets and suppressing distractors in past similar scenes can be memorized and transferred to new scenes under the facilitation of global scene characteristics. As shown in Fig. 6.9(m), the saliency maps generated by **MTLR** contain less noise than other top-down approaches. This also explains the reason that **MTLR** outperforms Navalpakkam07. In **MTLR**, only the experience on past similar scenes are transferred to the new scenes, while Navalpakkam07 infers a generalized solution for suppressing distractors. Often, such generalized experience may not apply to all scenes, particularly for those scenes with diversified contents. As shown in the last row of Fig. 6.9(k), the targets are suppressed while the distractors are mistakenly emphasized.

6.2.4.3   Comparisons with Ranking Models

In this experiment, we will test the performance of various learning-based ranking models on visual saliency computation and give explanations from the perspective of machine learning. The comparisons are made between **MTLR**, **Base-I** and two learning to rank approaches. This experiment adopts the same settings as the second experiment (e.g., the same training/testing sets, the same way to calculate the **ROC** curves). The **AUC** scores of various ranking models are shown in Table 6.2. The **ROC** curves are illustrated in Fig. 6.10 and some representative results are given in Fig. 6.11.

From Table 6.1 and Table 6.2, we can see that most ranking models outperforms the typical bottom-up and top-down visual saliency models. Compared with the learning-based top-down saliency models, the main advantage of ranking models is that they can better focus on the features that distinguishes targets from distractors. Particularly, we can see that the scene-specific models, including **Base-I** and **MTLR**, outperform the unified models Freund03 and Joachims06. Generally speaking, a video is not simply a collection of randomized scenes. Successive video frames often have similar targets and

Table 6.2 Performance of various ranking models on **ORIG**. Copyright © IEEE. All rights reserved. Reprinted, with permission, from (Li et al, 2011).

Algorithm	**AUC**	IMP (%)
Freund03	0.735	10.2
Joachims06	0.716	13.6
Base-I	0.739	9.72
MTLR	**0.811**	

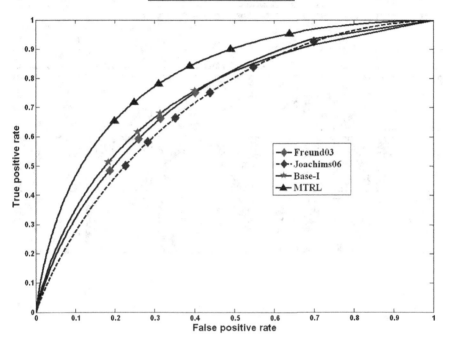

Fig. 6.10 ROC curves of **MTLR** and various ranking models. Copyright © IEEE. All rights reserved. Reprinted, with permission, from (Li et al, 2011).

distractors. By grouping similar training scenes into the same cluster, the model trained on this cluster can better apply to this kind of scenes. We can see that **Base-I** (**AUC**=0.739) outperforms Joachims06 (**AUC**=0.716) by simply grouping scenes and constructing cluster-specific models. However, the model trained on each cluster (particular for the cluster with limited number of scenes) often lacks the generalization ability and may become over-fitting due to the low diversity of training samples. In some scenes, **Base-I** with M=15 clusters may perform even worse than directly training **Base-I** (with M=1) on all scenes.

To avoid the over-fitting problem, **MTLR** adopts a multi-task learning framework to train multiple saliency models simultaneously with considering the inter-model correlations. With a properly designed multi-task learning algorithm, each model in **MTLR** is actually trained on the whole training set

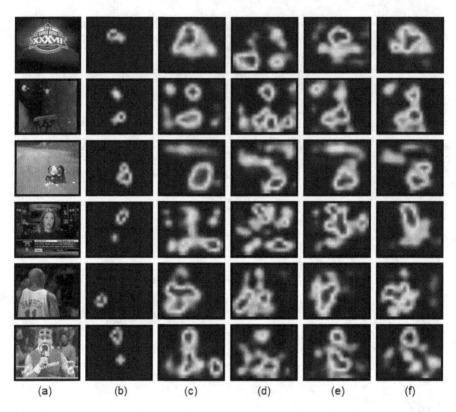

by emphasizing different subset of the training data. Therefore, the saliency models in **MTLR** have improved generalization ability and avoid over-fitting. Meanwhile, the model diversity is also guaranteed. Therefore, **MTLR** can outperform **Base-I** remarkably (as shown in Table 6.2 and Fig. 6.10).

6.2.4.4   Performance on Various Video Genres

In this experiment, we will test the performance of all the saliency and ranking models on different video contents. In the test, all the learning-based models are trained on $\mathbb{D}_{train}$ and tested on $\mathbb{D}_{test}$. Different from previous experiments, the performances are separately reported on the seven video genres of the **ORIG** dataset. The **AUC** scores of various approaches on different video genres are presented in Table 6.3. Moreover, the **ROC** curves of **MTLR** on different video contents are illustrated in Fig. 6.12.

Table 6.3 Performance of various approaches on different genres of **ORIG**. Copyright © IEEE. All rights reserved. Reprinted, with permission, from (Li et al, 2011).

	Outdoor	Video Game	Com'l	TV News	Sports	Talk Shows	Others
Itti98	0.559	0.584	0.530	0.531	0.536	0.552	0.569
Itti01	0.558	0.565	0.538	0.534	0.537	0.580	0.553
Itti05	0.615	0.640	0.593	0.587	0.577	0.679	0.647
Zhai06	0.662	0.632	0.684	0.656	0.590	0.598	0.632
Harel07	0.646	0.540	0.648	0.545	0.538	0.668	0.664
Hou07	0.684	0.661	0.684	0.660	0.607	0.699	0.696
Guo08	0.691	0.689	0.690	0.629	0.632	0.687	0.705
Kienzle07	0.541	0.537	0.519	0.509	0.518	0.516	0.555
Navalpakkam07	0.713	0.683	0.681	0.706	0.668	0.751	0.697
Peters07	0.670	0.724	0.631	0.686	0.687	0.680	0.670
Freund03	0.734	0.736	0.719	0.724	0.714	0.783	0.751
Joachims06	0.690	0.713	0.709	0.625	0.644	0.680	0.640
Base-I	**0.735**	**0.763**	**0.733**	**0.714**	**0.717**	**0.758**	**0.747**
MTLR	**0.824**	**0.833**	**0.806**	**0.773**	**0.783**	**0.818**	**0.808**

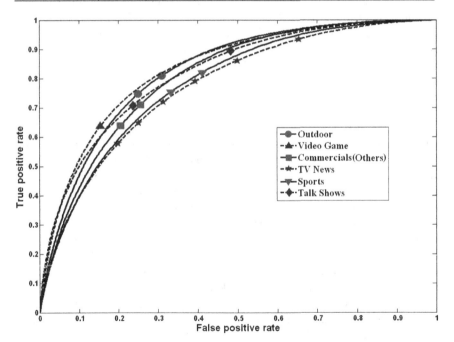

Fig. 6.12 ROC curves of **MTLR** on different video genres. Copyright © IEEE. All rights reserved. Reprinted, with permission, from (Li et al, 2011).

From Table 6.3, we find that most approaches perform the best in the "video game" genre. Often, the scenes in "video game" have obvious salient targets which can be easily distinguished from distractors. Actually, when playing the game, users often adjust the scene (e.g., change the view angle and character position) to ensure that the targets (e.g., the game characters) can easily pop-out. Moreover, the scenes in "video game" are often relatively simpler than the scenes in other video genres (e.g., the "TV news"), which will also assist the saliency computation.

In contrast to the "video game" genre, most algorithms achieve the worst performance on the "TV news" and "sports" genres. The main reason is that the scenes in "TV news" and "sports" are more complex than the scenes in other video genres. For example, a scene of "news" can have anchor, caption, scrolling text, logo and other contents, each of which can be a probable salient target (as shown in the fourth row of Fig. 6.9(a)). In such scenes with rich contents, it is difficult to distinguish targets from distractors only using the visual features. In some cases, the targets and distractors can be effectively separated only by using semantic clues. Therefore, incorporating the influences of various semantic clues (e.g., human face (Cerf et al, 2009; Ma et al, 2005) and camera motion (Abdollahian et al, 2008)) into visual saliency computation could be a challenging research direction.

6.3   Notes

In this Chapter, we propose a learning to rank approach for visual saliency computation. It has been proved that the pair-wise learning to rank framework can effectively learn the visual features that best distinguish targets from distractors. Moreover, the cluster-specific models can effectively pop-out targets and suppress distractors in various scenes. More importantly, a multi-task learning framework is adopted to infer multiple visual saliency models simultaneously. By an appropriate sharing of information across models, the performance of each model is greatly improved. From the results obtained so far, our approach outperforms several state-of-the-art bottom-up, top-down and learning to rank approaches remarkably.

One drawback of this approach is that it assumes all the training samples are accurately and sufficiently labeled. However, as discussed in Chap. 2, each video frame can only collect very limited fixations due to the short viewing period. In this case, the training samples are only sparsely labeled and it is inappropriate to treat all the unlabeled samples as negative. Since the learning to rank framework has been proved to be effective in this Chapter, we will discuss how to handle such label ambiguity problem in the ranking framework in next Chapter.

References

Abdollahian, G., Pizlo, Z., Delp, E.: A study on the effect of camera motion on human visual attention. In: Proceedings of the 15th IEEE International Conference on Image Processing (ICIP), pp. 693–696 (2008), doi:10.1109/ICIP.2008.4711849

Argyriou, A., Evgeniou, T., Pontil, M.: Multi-task feature learning. In: Advances in Neural Information Processing Systems (NIPS), pp. 41–48 (2007)

Boyd, S., Vandenberghe, L.: Convex Optimization. Cambridge University Press, Cambridge (2008)

Cerf, M., Harel, J., Einhauser, W., Koch, C.: Predicting human gaze using low-level saliency combined with face detection. In: Advances in Neural Information Processing Systems (NIPS), Vancouver, BC, Canada (2009)

Chun, M.M.: Contextual guidance of visual attention. In: Itti, L., Rees, G., Tsotsos, J. (eds.) Neurobiology of Attention, 1st edn., pp. 246–250. Elsevier Press, Amsterdam (2005)

Dempster, A., Laird, N., Rubin, D.: Maximum likelihood from incomplete data via the EM algorithm. Journal of the Royal Statistical Society, Series B 39(1), 1–38 (1977)

Evgeniou, T., Micchelli, C.A., Pontil, M.: Learning multiple tasks with kernel methods. Journal of Machine Learning Research 6, 615–637 (2005)

Freund, Y., Iyer, R., Schapire, R.E., Singer, Y.: An efficient boosting algorithm for combining preferences. Journal of Maching Learning Research 4, 933–969 (2003)

Guo, C., Ma, Q., Zhang, L.: Spatio-temporal saliency detection using phase spectrum of quaternion fourier transform. In: Preceedings of the IEEE Conference on Computer Vision and Pattern Recognition (CVPR), pp. 1–8 (2008), doi:10.1109/CVPR.2008.4587715

Harel, J., Koch, C., Perona, P.: Graph-based visual saliency. In: Advances in Neural Information Processing Systems (NIPS), pp. 545–552 (2007)

Hou, X., Zhang, L.: Saliency detection: A spectral residual approach. In: Preceedings of the IEEE Conference on Computer Vision and Pattern Recognition (CVPR), pp. 1–8 (2007), doi:10.1109/CVPR.2007.383267

Itti, L.: Crcns data sharing: Eye movements during free-viewing of natural videos. In: Collaborative Research in Computational Neuroscience Annual Meeting, Los Angeles, California (2008)

Itti, L., Baldi, P.: A principled approach to detecting surprising events in video. In: Preceedings of the IEEE Conference on Computer Vision and Pattern Recognition (CVPR), vol. 1, pp. 631–637 (2005), doi:10.1109/CVPR.2005.40

Itti, L., Koch, C.: Computational modeling of visual attention. Nature Review Neuroscience 2(3), 194–203 (2001)

Itti, L., Koch, C., Niebur, E.: A model of saliency-based visual attention for rapid scene analysis. IEEE Transactions on Pattern Analysis and Machine Intelligence 20(11), 1254–1259 (1998), doi:10.1109/34.730558

Jacob, L., Bach, F., Vert, J.P.: Clustered multi-task learning: A convex formulation. In: Advances in Neural Information Processing Systems (NIPS), pp. 745–752 (2009)

Joachims, T.: Training linear svms in linear time. In: Proceedings of the 12th ACM SIGKDD International Conference on Knowledge Discovery and Data Mining, KDD 2006, pp. 217–226. ACM, New York (2006), doi:10.1145/1150402.1150429

Kienzle, W., Wichmann, F.A., Scholkopf, B., Franz, M.O.: A nonparametric approach to bottom-up visual saliency. In: Advances in Neural Information Processing Systems (NIPS), pp. 689–696 (2007)

Li, J., Tian, Y., Huang, T., Gao, W.: Multi-task rank learning for visual saliency estimation. IEEE Transactions on Circuits and Systems for Video Technology 21(5),
623–636 (2011), doi:10.1109/TCSVT.2011.2129430

Ma, Y.F., Hua, X.S., Lu, L., Zhang, H.J.: A generic framework of user attention
model and its application in video summarization. IEEE Transactions on Multimedia 7(5), 907–919 (2005), doi:10.1109/TMM.2005.854410

Navalpakkam, V., Itti, L.: Search goal tunes visual features optimally. Neuron 53,
605–617 (2007)

Peters, R., Itti, L.: Beyond bottom-up: Incorporating task-dependent influences into
a computational model of spatial attention. In: Preeceedings of the IEEE Conference on Computer Vision and Pattern Recognition (CVPR), pp. 1–8 (2007),
doi:10.1109/CVPR.2007.383337

Pillonetto, G., Nicolao, G.D., Chierici, M., Cobelli, C.: Fast algorithms for nonparametric population modeling of large data sets. Automatica 45, 173–179 (2009)

Torralba, A.: Contextual influences on saliency. In: Itti, L., Rees, G., Tsotsos, J.
(eds.) Neurobiology of Attention, 1st edn., pp. 586–592. Elsevier Press, Amsterdam
(2005)

Treisman, A.M., Gelade, G.: A feature-integration theory of attention. Cognitive Psychology 12(1), 97–136 (1980)

Watrous, R.: Learning algorithms for connectionist networks: Applied gradient methods of nonlinear optimization, pp. 619–627, San Diego, CA, USA (1987)

Zhai, Y., Shah, M.: Visual attention detection in video sequences using spatiotemporal cues. In: Proceedings of the 14th Annual ACM International Conference
on Multimedia, MULTIMEDIA 2006, pp. 815–824. ACM, New York (2006),
doi:10.1145/1180639.1180824

Chapter 7
Removing Label Ambiguity in Training Saliency Model

This Chapter discusses the label ambiguities in training visual saliency models. Two approaches are presented to address this problem. The first approach proposes to embed the ambiguous data into the learning to rank framework in a cost-sensitive manner. The second approach adopts a multi-instance learning to rank framework to iteratively recover the correct sample labels and optimize the saliency model. These two approaches are proved to be effective in improving the performance of learning-based visual saliency computation.

7.1 Overview

To estimate visual saliency, many approaches aim to learn a "feature-saliency" mapping model by using the user data obtained from manually labeling activities or eye-tracking devices. These approaches often divide an image or video frame into a set of patches (e.g., macro-blocks or objects). After that, each patch is assigned a label which can be obtained from the user data to indicate whether it is salient or not. By representing each patch as a high-dimensional feature vector and utilizing these labels, the learning process can be described as inferring a mapping function that can transform high-dimensional feature vectors into real saliency values. For example, Kienzle et al (2007) presented a non-parametric saliency model to map intensity values to saliency values by using the Support Vector Machine. Navalpakkam and Itti (2007) adopted a learning-based algorithm to pop-out the targets and suppress the distractors through maximizing the signal-noise-ratio. Generally speaking, these learning-based approaches can achieve promising results since they can transfer the user's experience from the training data to new cases in a supervised manner.

However, when computing visual saliency on existing benchmarks, two kinds of label ambiguities may arise due to the inaccurate and inadequate user data. As shown in the upper part of Fig. 7.1, salient objects are often manually labeled by rectangles (e.g., Liu et al, 2007) since it is really time-consuming to label the accurate boundaries for salient objects with irregular shapes. In the training process, these inaccurate user data may generate many pseudo targets (i.e., real distractors wrongly labeled as targets). In contrast, when labeling images with eye-tracking devices, some salient image patches may be wrongly labeled or remain unlabeled due to the inadequate viewing time or the gap between covert and overt visual attention, particularly for those images with multiple or large salient objects (as shown in the lower part of Fig. 7.1). Correspondingly, many pseudo distractors (i.e., real targets wrongly labeled as distractors) are also used as the training data. Without considering such label ambiguities, these pseudo targets/distractors will degrade the learned model, making the estimated visual saliency no longer reliable.

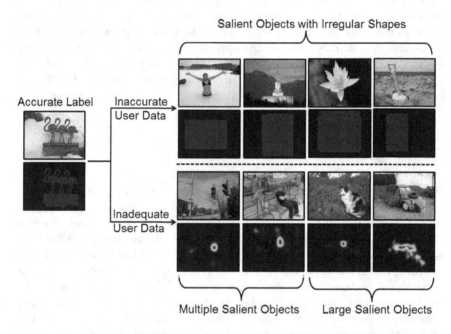

Fig. 7.1 Label ambiguity may arise from inaccurate or inadequate user data. Copyright © IEEE. All rights reserved. Reprinted, with permission, from (Li et al, 2012).

This Chapter discusses how to remove the label ambiguities in training visual saliency models. Two approaches are presented to address this problem in image and video benchmarks, respectively. The first approach proposes to embed the ambiguous training instances into a learning to rank framework in a cost-sensitive manner. The second approach adopts a multi-instance learning to rank framework to iteratively recover the correct labels of training instances and optimize the saliency model. In addition, the first approach focuses on visual saliency in video while the second emphasizes image saliency.

7.2   A Cost-Sensitive Learning to Rank Approach

To remove the label ambiguities from the fixations recorded on video, this Section presents a cost-sensitive learning to rank approach (Li et al, 2010). In this approach, the ambiguous training instances are incorporated into a pair-wise learning to rank framework. In this framework, we avoid the explicit extraction of positive and negative samples by directly integrating both the positive and unlabeled data into the optimization objective in a cost-sensitive manner. Extensive experiments demonstrate that this approach outperforms several state-of-the-art bottom-up and top-down approaches in visual saliency computation.

7.2.1   The Approach

Often, users intend to search the desired targets under the facilitation of experience derived from past similar scenes. In this process, a requisite step is to identify the search priority of each location. Such priority is tightly related to visual saliency and can be derived from local visual attributes and pair-wise contexts. Therefore, we can formulate visual saliency computation as a learning to rank problem to estimate the searching priority of each location.

Here we first describe how to extract the local visual attributes. In general, a scene can be expressed as a conjugate of information flows from multiple visual feature channels. Among these channels, some are the probable sources of preattentive guidance and visual saliency is related to local contrasts in these preattentive channels. Here we compute the local contrasts from 12 typical preattentive channels in 6 scales, including intensity (6), color opponencies (12), orientations (24), flickers (6) and motion energies (24). In total 72 local contrasts are obtained. Here we use a vector $\mathbf{x}_{kn}$ with 72 components to characterize the local visual attributes of the n-th location $\mathbf{B}_{kn}$ (e.g., 16×16 block) in the k-th scene.

Using these features, we then present the cost-sensitive learning to rank approach for visual saliency computation. In the learning process, an important issue is to train a ranking function $\phi(\mathbf{x})$ using the ground truth saliency g_{km}. For two locations $\mathbf{B}_{km}$ and $\mathbf{B}_{kn}$, $\phi(\mathbf{x}_{km}) > \phi(\mathbf{x}_{kn})$ indicates that $\mathbf{B}_{km}$ ranks higher than $\mathbf{B}_{kn}$ and maintains a higher saliency. However, the user data often contain only sparse positive samples. As shown in Fig. 7.2(a), the eye traces can only reveal parts of the salient target, while most locations remain unlabeled. To utilize the unlabeled data, we derive g_{km} by considering

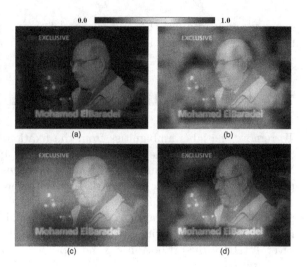

the visual similarity and the spatial correlation between $\mathbf{B}_{km}$ and the labeled
positive samples. Let e_{kn} be the event that $\mathbf{B}_{kn}$ is a labeled positive sample,
the visual similarity v_{km} can be calculated as:

$$v_{km} = \max_{n \in \{1,\ldots,N\}} \left(\frac{[e_{kn}]_{\mathbf{I}} \cdot \mathbf{x}_{km}^{\mathbf{T}} \mathbf{x}_{kn}}{\|\mathbf{x}_{km}\| \cdot \|\mathbf{x}_{kn}\|} \right), \tag{7.1}$$

where $[e_{kn}]_{\mathbf{I}} = 1$ if e_{kn} holds, otherwise $[e_{kn}]_{\mathbf{I}} = 0$. N is the total number of
blocks in a scene. As shown in Fig. 7.2(c), such visual similarities can pop-out
the locations that are similar to the positive samples. Moreover, the spatial
correlation r_{km} is computed as:

$$r_{km} = \max_{n \in \{1,\ldots,N\}} \exp \left(- \left(\frac{[e_{kn}]_{\mathbf{I}} \cdot d_{mn}}{d_k} \right) \right), \tag{7.2}$$

where d_{mn} is the Euclidean distance between the locations $\mathbf{B}_{km}$ and $\mathbf{B}_{kn}$,
while d_k corresponds to the diagonal distance of the k-th scene. As shown
in Fig. 7.2(c), such spatial correlations can pop-out the remainder of the
salient target. After that, we normalize the visual similarities and the spatial
correlations into [0,1] and derive g_{km} by setting:

$$g_{km} = v_{km} \cdot r_{km}. \tag{7.3}$$

As shown in Fig. 7.2(d), the formulation in (7.3) will only assign high
saliency values to the locations that are adjacent and similar to the labeled
positive samples. In the training process, however, it is often difficult to directly determine the label for each sample, especially for the one with medium
saliency (e.g., around 0.5). Moreover, visual saliency computation mainly focuses on distinguishing targets from distractors and the correlations between

target pairs or between distractor pairs should be considered with low priority. Therefore, we integrate all the positive and unlabeled data (with estimated ground truth saliency) into a learning to rank framework in a cost-sensitive manner. Here we adopt a ranking function $\phi(\mathbf{x}) = \omega^{\mathbf{T}}\mathbf{x}$ to combine the input features with linear weights. Given the local visual attributes and ground-truth saliency for each location of the training scenes, the empirical loss can be defined as:

$$\mathcal{L}(\omega) = \sum_k \sum_{m \neq n}^{N} [g_{km} - g_{kn}]_+ \cdot [\omega^{\mathbf{T}}\mathbf{x}_{km} \leq \omega^{\mathbf{T}}\mathbf{x}_{kn}]_{\mathbf{I}}, \qquad (7.4)$$

where $[x]_+ = \max(0, x)$. We can see that there will be a loss if the ranking function gives predictions contrary to the ground-truth saliencies (i.e., $\phi(\mathbf{x}_{km}) \leq \phi(\mathbf{x}_{kn})$ when $g_{km} > g_{kn}$). Moreover, the loss function emphasizes the correlations between targets and distractors since the central issue in visual saliency computation is to distinguish targets from distractors. That is, the cost of erroneously ranking a target-distractor pair (i.e., $g_{km} - g_{kn} \to 1$) is much bigger than that of mistakenly predicting the ranks between target pairs or between distractor pairs (i.e., $g_{km} - g_{kn} \to 0$). Thus it is cost-sensitive by differentiating target-distractor pairs in the framework.

Generally speaking, minimizing the loss function (7.4) will not only pop-out the target using the local visual attributes, but also suppress the distractors by considering the influences of pair-wise correlations. However, it is difficult to directly minimize such a binary loss function. Toward this end, a feasible solution is to find an upper bound of (7.4) and minimize the upper bound instead. From the definition of the loss function, we can see that:

$$[\omega^{\mathbf{T}}\mathbf{x}_{km} \leq \omega^{\mathbf{T}}\mathbf{x}_{kn}]_{\mathbf{I}} \leq e^{\omega^{\mathbf{T}}(\mathbf{x}_{kn} - \mathbf{x}_{km})}. \qquad (7.5)$$

By incorporating (7.5) into (7.4), the optimization objective can be rewritten as:

$$\min_{\omega} \sum_k \sum_{m \neq n}^{N} [g_{km} - g_{kn}]_+ \cdot e^{\omega^{\mathbf{T}}(\mathbf{x}_{kn} - \mathbf{x}_{km})}. \qquad (7.6)$$

Note that (7.6) contains only exponential terms with linear positive weights. Thus the objective function is convex and the global optimum can be reached using gradient-based method:

$$\triangle \omega \propto \sum_k \sum_{m \neq n}^{N} [g_{km} - g_{kn}]_+ \cdot (\mathbf{x}_{kn} - \mathbf{x}_{km}) \cdot e^{\omega^{\mathbf{T}}(\mathbf{x}_{kn} - \mathbf{x}_{km})}. \qquad (7.7)$$

With the derived ranking function, a rank $c_{km} \in \{1, ..., N\}$ is assigned to each block $\mathbf{B}_{km}$. To get the visual saliency map, we empirically turn this rank into a saliency value $\left(\frac{N - c_{km}}{N}\right)^{\beta}$, where $\beta > 0$ is a constant to pop-out the probable salient targets. For larger β (empirically set to 3 in this study), the locations other than the most salient location can be suppressed more effectively. Moreover, we convolve the saliency map with a Gaussian kernel ($\sigma = 5$) to ensure that the entire salient object can be detected. For convenience, the saliency values are normalized into [0,1].

7.2.2   *Experiments*

To evaluate the feasibility of this cost-sensitive learning to rank approach for visual saliency computation, we adopt one of the video dataset proposed in (Itti, 2008) with eye traces of 8 subjects. The dataset, denoted as **ORIG**, consists of over 46,000 video frames in 50 video clips (25 minutes). When watching each video clip, the eye traces of 4-6 subjects were recorded using a 240HZ eye-tracker. In the experiment, we randomly divide the dataset into training/validation/test sets for 10 times. For each division, we adopt 10 state-of-the-art approaches for comparison. In general, these approaches can be grouped into two categories:

1. **Bottom-Up approaches**, including seven location-based bottom-up models such as Itti98 (Itti et al, 1998), Itti01 (Itti and Koch, 2001), Itti05 (Itti and Baldi, 2005), Zhai06 (Zhai and Shah, 2006), Harel07 (Harel et al, 2007), Hou07 (Hou and Zhang, 2007) and Guo08 (Guo et al, 2008);
2. **Top-Down approaches**, including Peters07 (Peters and Itti, 2007), Kienzle07 (Kienzle et al, 2007) and Navalpakkam07 (Navalpakkam and Itti, 2007).

These approaches will be compared with the proposed cost-sensitive learning to rank approach, which is denoted as **CSLR**. Among these approaches, Itti98, Itti01, Itti05 and Navalpakkam07 adopt the same local visual features as in this approach. For fair comparison, the model of Kienzle07 is also trained with these features other than the local intensities. In the experiments, the parameters for the learning-based approaches are optimized on the validation set. In the experiments, we use the Area Under the ROC Curve (**AUC**) to evaluate the overall performance. The **AUC** scores are shown in Table 7.1[1] and some representative results are given in Fig. 7.3.

From Table 7.1, we can see that **CSLR** outperforms all the other ten approaches. As shown in Fig. 7.3(c)-(d), Itti98 and Itti01 only maintain the most salient locations using the "winner-take-all" competition. Thus they yield low **AUC** scores since the competition may not only suppress the distractors but also inhibit the targets. Other bottom-up approaches perform much better by integrating the irregularities in the spatiotemporal domain

---

[1] The **AUC** scores reported in Tables 5.2, 6.1 and 7.1 may change slightly due to different experimental settings (e.g., sizes of Gaussian kernels used in convolution, resolutions of saliency maps and settings of training/validation/testing sets).

Table 7.1 Performance of various approaches on **ORIG**. Copyright © IEEE. All rights reserved. Reprinted, with permission, from (Li et al, 2010).

Algorithm		**AUC**	Improvement (%)
Bottom-up	Itti98	0.571	35.6
	Itti01	0.564	37.2
	Itti05	0.643	20.4
	Hou07	0.680	13.8
	Guo08	0.697	11.0
	Harel07	0.590	31.2
	Zhai06	0.633	22.3
Top-down	Peters07	0.630	22.9
	Kienzle07	0.533	45.2
	Navalpakkam07	0.701	10.4
	CSLR	**0.774**	

(e.g., Itti05, Zhai06 and Harel07), in the amplitude spectrum (e.g., Hou07) or in the phase spectrum (e.g., Guo08). However, the experience on past scenes can often provide effective clues for suppressing the distractors, which are not considered by these approaches. Thus the saliency maps generated by these approaches often contain much noise.

Compared with the bottom-up approaches, the top-down approaches (e.g., Peters07, Kienzle07 and Navalpakkam07) do not demonstrate much improvement due to their inappropriate strategies on sampling negative training samples. For example, Peters07 and Navalpakkam07 treated all the locations that have received no fixations as negative samples (i.e., distractors), while Kienzle07 generated the negative samples by using the same coordinates of eye fixations, but on different scenes. Therefore, many positive samples in the unlabeled data will be assigned to wrong labels, which will greatly degrade the overall performance. As shown in Fig. 7.3(j) - (l), these approaches can only recover parts of the targets, while the other parts are often suppressed as distractors.

Compared with these approaches, **CSLR** can accurately locate the salient target while suppressing the distractors. Particularly, **CSLR** can pop-out the entire salient object, other than the most salient points. As shown in the second row of Fig. 7.3(m), **CSLR** can pop-out the whole caption line and the car from a complex scene. Generally speaking, the success of **CSLR** is mainly due to two reasons. First, the positive and unlabeled data is directly exploited by the optimization objective in a cost-sensitive manner, which provides effective clues for locating the entire targets. Second, the learning to rank framework focuses not only on the local visual attributes but also on the pair-wise correlations. In the learning process, the local visual attributes assist to pop-out the targets, while pair-wise correlations can help to suppress the distractors.

We also design an experiment to illustrate why the cost-sensitive formulation is necessary. In one experiment, we set a threshold T_D to select only

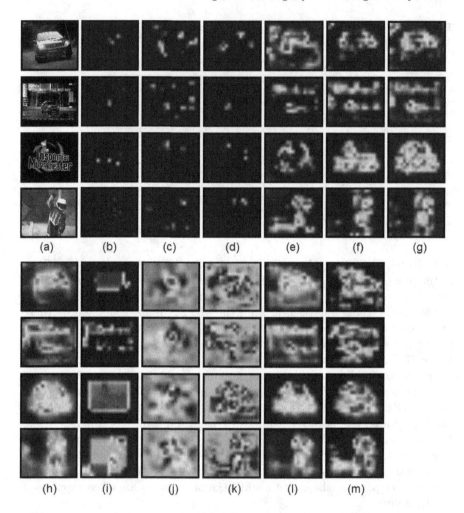

Fig. 7.3 Representative results obtained by various saliency models. (a) Original frames; (b) eye fixation maps; (c) Itti98; (d) Itti01; (e) Itti05; (f) Hou07; (g) Guo08; (h) Harel07; (i) Zhai06; (j) Peters07; (k) Kienzle07; (l) Navalpakkam07; (m) **CSLR**. Copyright © IEEE. All rights reserved. Reprinted, with permission, from (Li et al, 2010).

the sample pairs whose saliency differences are larger than T_D into the training process. In this process, the selected sample pairs are treated with equal weights (i.e., the term $[g_{km} - g_{kn}]_+$ in (7.4) is fixed to 1). First, we set $T_D \approx 1$ and the **AUC** score reaches 0.743. This indicates that the learning to rank framework itself can improve the overall performance, even only using the most reliable samples. Then we gradually decrease T_D and the **AUC** reaches its maximal (0.775) when $T_D \approx 0.6$, and then decrease to 0.766 when $T_D = 0$. The reason is that the correlations between targets and distractors (i.e., sample pairs with high saliency differences) can provide effective clues for selecting the discriminative features for the ranking function. However, correlations between targets or between distractors (i.e., sample pairs with low saliency differences) may bias the selection process to the wrong visual attributes, thus decreasing the **AUC** score. This implies that penalizing correlations between targets and between distractors in a cost-sensitive manner is helpful in visual saliency computation.

From the results obtained so far, we can see that a simple learning to rank framework can achieve impressive performance by simultaneously considering the influences of both the local visual attributes and pair-wise contexts. Moreover, By directly integrating the positive and unlabeled data into the optimization objective in a cost-sensitive manner, the learned saliency model can outperform other learning-based approaches by focusing on the correlations between reliable targets and distractors. One drawback of this approach is it only considers the case of training with sparse positive samples. In the next Section, we will extend this framework to a general framework which aims to train saliency models with probable pseudo targets and pseudo distractors.

7.3 A Multi-instance Learning to Rank Approach

To effectively cope with these weakly labeled data with different types of label ambiguities, this Section propose a multi-instance learning to rank approach for visual saliency computation. Different from traditional learning algorithms, the label ambiguities are also considered in the multi-instance learning to rank approach. Generally, multi-instance learning (MIL) methods aim to learn the models from the data with label ambiguities (referred to as weakly labeled data), in which the training instances are from the positive and negative bags. In the traditional multi-instance learning constraints, each positive bag is assumed to contain at least one (e.g., Andrews et al, 2003) or sparse (e.g., Bunescu and Mooney, 2007) positive instances and all the instances in each negative bag are assumed to be negative.

Inspired by this idea, the actual label of each image patch when training visual saliency model is also considered to be unknown. Instead, the minimal number of real targets in each positive bag and real distractors in each negative bag are incorporated as constraints (as suggested in Andrews et al, 2003). Moreover, this approach further adopts an ordinal regression framework by only utilizing the pair-wise correlations between image patches from positive and negative bags. These correlations (one patch is more/equally/less salient than another patch), can be used to train a ranking model as well as to relabel the patches. By iteratively refining the ranking model and updating the patch labels, this approach can effectively cope with such label ambiguities in the training data. Finally, the learned ranking model can be used for visual saliency computation by popping-out real targets and suppressing real distractors. To demonstrate the effectiveness of this approach, extensive experiments are carried out on two public image datasets which contain rich pseudo targets and rich pseudo distractors, respectively. Experimental results show that this approach outperforms several bottom-up and top-down methods remarkably.

7.3.1 Problem Formulation

Generally speaking, visual saliency can be viewed as an indicator for the relative importance of various objects in an image. However, the absolute saliency values of the same object may vary remarkably in different images. Therefore, a direct "feature-saliency" mapping function from local visual attributes to visual saliency values may not exist. Toward this end, we assume that *visual saliency should be jointly determined by intrinsic visual attributes and spatial/spatiotemporal visual contexts.*

Following this assumption, the influences of intrinsic visual attributes and visual contexts should be considered simultaneously in the learning process. Actually, the user data obtained either from manually labeling activities or eye-tracking devices can only reveal the correlations between salient objects and their contexts (i.e., one object is more/equally/less salient than another object). When training visual saliency models, such relative saliency correlations should be emphasized to distinguish salient targets from distractors. Therefore, it is more reasonable to represent a visual saliency model with a ranking function. Thus the learning objective is to infer a ranking function that is optimized to handle such relative saliency correlations in the training data.

When training the visual saliency model, there often exist many pseudo targets/distractors among the training data. Usually, such label ambiguity may arise when the user data are *inaccurate* or *inadequate*. To effectively cope with such pseudo targets/distractors, we propose a multi-instance learning to rank approach for visual saliency computation. In this approach, such pseudo targets/distractors can be relabeled with respect to their mutual correlations and a ranking function is optimized to distinguish **real targets** from **real distractors**. For the sake of simplicity, we first introduce some notations:

- Let E_k^+ and E_k^- be two instance bags which contain the image patches (i.e., 16×16 macro-blocks in this work) from image I_k with positive and negative user labels, respectively. We also denote $|E_k^+|$ and $|E_k^-|$ as the numbers of instances in these two bags and let $N_k = |E_k^+| + |E_k^-|$ be the total number of image patches in image I_k.
- $\mathbf{x}_{ki}$ is the feature vector that represents a patch e_{ki}, while y_{ki} and g_{ki} are its real label and user label. That is, $g_{ki} = +1$ for $e_{ki} \in E_k^+$ and $g_{ki} = -1$

for $e_{ki} \in E_k^-$. In contrast, $y_{ki} = +1$ if e_{ki} is a real target and $y_{ki} = -1$ if e_{ki} is a real distractor.

- v_{kij} represents an ordered patch pair $\overrightarrow{(e_{ki}, e_{kj})}$ and $\mathcal{V} = \{v_{kij} | y_{ki} = +1, y_{kj} = -1\}$ denotes the set of all such target-distractor pairs in the training data.
- $\phi(\mathbf{x}) = \mathbf{w}^\mathrm{T}\mathbf{x}$ is a ranking function with which image patch e_{ki} is more salient than e_{kj} if $\phi(\mathbf{x}_{ki}) > \phi(\mathbf{x}_{kj})$.

From the above notations, we can see that the training data can be represented by a set of instance bags $\mathcal{E} = \{E_k^+, E_k^-\}$. Consequently, the objective of the learning process can be described as inferring the real labels $\{y_{ki}\}$ for the instances in $\mathcal{E}$ and optimizing a ranking function $\phi(\mathbf{x})$ that can best distinguish real targets from real distractors.

We first focus on the image dataset in which salient objects are manually labeled with rectangles. In this case, image patches in the labeled rectangles may be real targets and pseudo targets, while image patches outside of the rectangles are real distractors. Suppose that there exists at least one real target in each positive bag, the optimization problem can be formulated as:

$$
\begin{aligned}
&\min_{\{y_{ki}\}} \min_{\mathbf{w}, \{\xi_v\}} \frac{1}{2}\|\mathbf{w}\|_2^2 + C\sum_{v \in \mathcal{V}} \xi_v, \\
&s.t. \ \ \forall v \in \mathcal{V}, \ \mathbf{w}^\mathrm{T}(\mathbf{x}_{ki} - \mathbf{x}_{kj}) \geq 1 - \xi_v, \xi_v \geq 0, \\
&\quad\quad \forall k, \ \sum_{e_{ki} \in E_k^+} \frac{y_{ki} + 1}{2} \geq 1, \\
&\quad\quad \forall k, y_{ki} = -1, \forall e_{ki} \in E_k^-, \\
&\quad\quad \forall k, i, y_{ki} \in \{-1, 1\},
\end{aligned}
\tag{7.8}
$$

where ξ_v is a slack variable that measures the degree of misclassification for the instance $\mathbf{x}_{ki} - \mathbf{x}_{kj}$. Note that here the second constraint indicates that there exists at least one real target in the positive bag E_k^+, while the third constraint indicates that all the instances in the negative bag E_k^- are real distractors.

Similarly, most of the users' eye traces can accurately locate the salient objects. However, some unattended salient patches may be wrongly labeled as pseudo distractors, while some attended patches may also be pseudo targets due to the gap between overt and covert attention. In this case, a positive bag E_k^+ often contains rich real targets while a negative bag E_k^- may also contain pseudo distractors. Considering this assumption on the eye-fixation data and (7.8), we now introduce a more general formulation by using two sets of parameters $\{\rho_k^+\}$ and $\{\rho_k^-\}$. Here ρ_k^+ indicates the minimum percentage of real targets in E_k^+, while ρ_k^- indicates the minimum percentage of real distractors in E_k^-. Given these two sets of parameters, the problem can be reformulated in a general form:

$$\min_{\{y_{ki}\}} \min_{\mathbf{w},\{\xi_v\}} \frac{1}{2}\|\mathbf{w}\|_2^2 + C \sum_{v \in \mathcal{V}} \xi_v,$$

$$s.t. \ \forall v \in \mathcal{V}, \ \mathbf{w}^T (\mathbf{x}_{ki} - \mathbf{x}_{kj}) \geq 1 - \xi_v, \xi_v \geq 0,$$

$$\forall k, \sum_{e_{ki} \in E_k^+} \frac{y_{ki} + 1}{2} \geq \rho_k^+ |E_k^+|, \tag{7.9}$$

$$\forall k, \sum_{e_{ki} \in E_k^-} \frac{1 - y_{ki}}{2} \geq \rho_k^- |E_k^-|,$$

$$\forall k, i, y_{ki} \in \{-1, 1\}.$$

Note that here the second and the third constraints indicate that there are at least $\rho_k^+ |E_k^+|$ real targets in the positive bag E_k^+ and $\rho_k^- |E_k^-|$ real distractors in the negative bag E_k^-. Generally, the parameters ρ_k^+ and ρ_k^- can be empirically predefined (e.g., through subjective observation):

- When salient objects are manually labeled by one or several rectangles, $\rho_k^+ = \frac{1}{|E_k^+|}$ and $\rho_k^- = 1$.
- When salient objects are located and tracked by human eye fixations, $\rho_k^+ \approx 1$ and $\rho_k^- = \frac{1}{|E_k^-|}$.
- When salient objects are segmented from the background with accurate contours, $\rho_k^+ = 1$ and $\rho_k^- = 1$.

7.3.2   The Learning Algorithm

In this Section, we will address four key issues for solving the proposed multi-instance learning to rank problem. First, we will extract the local visual attributes that are used to characterize an image patch. With these visual attributes, the redundant instances are then removed to improve the computational efficiency. After that, a learning algorithm is presented to cope with the label ambiguities in the training data and optimize the ranking model. Finally, we describe the details for estimating visual saliency of the testing image with this ranking model.

7.3.2.1   Extracting Local Visual Attributes

We now describe how to extract the local visual attributes $\mathbf{x}_{ki}$ to characterize an image patch e_{ki}. Here we extract two kinds of features to construct $\mathbf{x}_{ki}$, including "center-surround" local contrasts and location features. The "center-surround" local contrasts are widely used in many approaches and we adopt the algorithm in (Itti et al, 1998) to extract them. Specifically, a Gaussian pyramid is used to extract 42 local "center-surround" contrasts from 6 scales and 7 preattentive features (intensity, red/green and blue/yellow opponencies and four orientations). Generally, salient objects have strong responses in one or several local contrasts since they can be easily popped-out from their surroundings. Note that here these raw local contrasts are directly used to construct $\mathbf{x}_{ki}$ and we do not conduct max-normalization steps on each image.

Beyond the local contrast features, the location feature is also very important for visual saliency computation. According to (Tseng et al, 2009), it is demonstrated that user often prefers to observe the image centers with high priorities. This implies that salient objects appear around the image centers with high probabilities. Therefore, the location feature is computed as the distance from an image patch to the image center, which is further normalized using the diagonal distance of the image.

By combining the 42 local contrast features and location feature, the local visual attributes of an image patch e_{ki} are represented by a 43-dimensional column vector $\mathbf{x}_{ki}$. Note that each component of $\mathbf{x}_{ki}$ is further normalized into the range [0,1] in this work.

7.3.2.2   Removing Redundant Instances

Given the local visual attributes, we can then solve the optimization problem in (7.9). However, we can see that the pair-wise correlations between targets and distractors are considered in (7.9). In this case, $O(N)$ image patches can generate $O(N^2)$ pair-wise correlations, making the optimization process computationally expensive.

To improve the computational efficiency, removing the redundant training data is a feasible solution. Actually, there exist rich redundancies in the training data since many patches in the same image are often similar to each other (as shown in Fig. 7.4). For these similar image patches, we can replace them with only one representative instance (i.e., an exemplar). Consequently, the number of pair-wise correlations can decrease quadratically and the overall computational cost can be greatly reduced.

An important issue for redundancy reduction is how to select the most representative instances from the training data. A straightforward solution is to randomly choose an initial subset of instances as exemplars and then iteratively refining them with K-means. However, the results of this solution heavily depend on the choice of the initial subset of instances. Considering that it is difficult to determine the optimal number of exemplars to represent all the training data, we use the affinity propagation algorithm (Frey and Dueck, 2007) to automatically select a proper portion of instances as exemplars. In this process, all instances are simultaneously considered as potential exemplars. By iteratively exchanging real-valued messages between instances with respect to their similarities, a set of exemplars and the corresponding

Fig. 7.4 Exemplar selection. 1st row: images; 2nd row: labeled rectangles covering the entire salient objects; 3rd row: selected exemplars. Copyright © IEEE. All rights reserved. Reprinted, with permission, from (Li et al, 2012).

clusters gradually emerge. In this work, the similarity of two image patches is defined as the L_1 distance between their local visual attributes.

After removing the redundancies from the training data, only a limited number of exemplars are selected for further processing. For the sake of simplicity, we denote $\widehat{E}_k^+$ and $\widehat{E}_k^-$ as two instance bags which only contain the exemplars from image I_k with positive and negative user labels, respectively. As shown in the third row of Fig. 7.4, such exemplars are much less than the original instances and they can well represent the training data. Generally speaking, there are two main advantages to remove the redundant instances from the training data. First, the computational efficiency for solving the optimization problem in (7.9) can be greatly improved by replacing a large amount of training data with only the representative instances. Second, removing redundant training samples can also alleviate the probable over-fitting issue in the learning process. In some cases, salient targets and background regions may both contain many near-duplicate image patches. Consequently, the numerous correlations between these patches may overwhelm other unique or rare correlations (e.g., the correlations between small salient objects and small distractors) and lead to the over-fitting problem.

7.3.2.3 Learning

From (7.9), we can see that this formulation leads to a mixed integer programming problem. That is, the integer label of each exemplar as well as the ranking function to distinguish real targets from real distractors should be optimized simultaneously. However, it is difficult to directly solve this problem since the objective function is not convex. Therefore, we have to update the exemplar labels and optimize the ranking function iteratively. Toward this end, we use the EM algorithm (Dempster et al, 1977) to iteratively solve the problem and ensure the convergence. As shown in Fig. 7.5, the optimization process consists of two major steps: 1) fixing all exemplar labels and optimizing a ranking function $\phi(\mathbf{x})$ based on the pair-wise correlations; and 2) using this ranking function to update the labels for exemplars. By iteratively updating the exemplar labels, generating exemplar pairs and training the ranking function, the visual saliency model (i.e., the ranking function $\phi(\mathbf{x})$) can be gradually refined by focusing on the correlations between real targets and real distractors.

The details of the learning algorithm are listed as follows. In the initialization step, the exemplar label y_{ki} is initialized as the user label g_{ki} and an optimal ranking function is then trained by minimizing (7.9). After that, we can iteratively update $\{y_{ki}\}$ and optimize $\mathbf{w}$ in two steps:

Update $\{y_{ki}\}$: Given the ranking function $\phi(\mathbf{x}) = \mathbf{w}^{\mathrm{T}}\mathbf{x}$, we aim to update the exemplar labels $\{y_{ki}\}$ in this step. Observing we have two constraints $\mathbf{w}^{\mathrm{T}}(\mathbf{x}_{ki} - \mathbf{x}_{kj}) \geq 1 - \xi_v$ and $\xi_v \geq 0$ on ξ_v in (7.9), we can set $\xi_v = \max(0, 1 - \mathbf{w}^{\mathrm{T}}(\mathbf{x}_{ki} - \mathbf{x}_{kj}))$ and rewrite the objective function in (7.9) as:

$$\min_{\{y_{ki}\}} \sum_k \sum_{\substack{y_{ki}=+1 \\ y_{kj}=-1}} \max(0, 1 - \mathbf{w}^{\mathrm{T}}(\mathbf{x}_{ki} - \mathbf{x}_{kj})),$$

$$s.t. \forall k, \quad \sum_{e_{ki} \in \widehat{E}_k^+} \frac{y_{ki} + 1}{2} \geq \rho_k^+ |\widehat{E}_k^+|, \tag{7.10}$$

$$\forall k, \quad \sum_{e_{ki} \in \widehat{E}_k^-} \frac{1 - y_{ki}}{2} \geq \rho_k^- |\widehat{E}_k^-|,$$

$$\forall k, i, y_{ki} \in \{-1, 1\}.$$

For the sake of simplification, we rewrite this objective function in a matrix form. Let $\mathbf{y}_k$ and $\mathbf{g}_k$ be two l_k-dimensional column vectors with the ith components equal to y_{ki} and g_{ki}, respectively. Here $l_k = |\widehat{E}_k^+| + |\widehat{E}_k^-|$ is the total number of exemplars in image I_k, which is usually much less than the total number of image patches N_k. Meanwhile, let $\mathbf{1}$ be a l_k-dimensional column vector with all elements equal to one and $\mathbf{M}_k$ be a $l_k \times l_k$ matrix with $(\mathbf{M}_k)_{ij} = \max(0, 1 - \mathbf{w}^{\mathrm{T}}(\mathbf{x}_{ki} - \mathbf{x}_{kj}))/4$. Therefore, the optimization objective can be rewritten as:

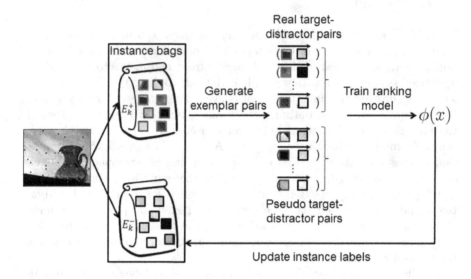

Fig. 7.5 System framework of the approach in (Li et al, 2012). Copyright © IEEE. All rights reserved. Reprinted, with permission, from (Li et al, 2012).

$$\min_{\{\mathbf{y}_k\}} \sum_k (\mathbf{1} + \mathbf{y}_k)^\mathrm{T} \mathbf{M}_k (\mathbf{1} - \mathbf{y}_k),$$

$$s.t. \forall k, (\mathbf{1} + \mathbf{g}_k)^\mathrm{T}(\mathbf{1} + \mathbf{y}_k) \geq 4\rho_k^+ |\widehat{E}_k^+|, \qquad (7.11)$$

$$\forall k, (\mathbf{1} - \mathbf{g}_k)^\mathrm{T}(\mathbf{1} - \mathbf{y}_k) \geq 4\rho_k^- |\widehat{E}_k^-|,$$

$$\forall k, i, y_{ki} \in \{-1, 1\}.$$

where $\mathbf{1} + \mathbf{g}_k$ (*resp.*, $\mathbf{1} + \mathbf{y}_k$) is one vector in which the entries are equal to 2 when corresponding to the instances with positive user labels (*resp.*, real labels), or 0, otherwise. We can see that the optimization problem in (7.11) is NP-hard and it is computationally expensive to determine the optimal binary label for each exemplar. However, we observe that the objective function only contains quadratic terms of $\{y_{ki}\}$ and the first two sets of constraints on $\{y_{ki}\}$ are linear. To solve this problem, we first perform a relaxation on $\{y_{ki}\}$ by replacing the third set of constraints on $\{y_{ki}\}$ as:

$$\forall k, i, -1 \leq y_{ki} \leq 1. \qquad (7.12)$$

With this relaxation, the optimization problem in (7.13) contains only quadratic terms with linear constraints. Thus we can solve each $\mathbf{y}_k$ separately:

$$\min_{\mathbf{y}_k} \ (1 + \mathbf{y}_k)^{\mathrm{T}} \mathbf{M}_k (1 - \mathbf{y}_k),$$

$$s.t. (1 + \mathbf{g}_k)^{\mathrm{T}} (1 + \mathbf{y}_k) \geq 4\rho_k^+ |\widehat{E}_k^+|,$$

$$(1 - \mathbf{g}_k)^{\mathrm{T}} (1 - \mathbf{y}_k) \geq 4\rho_k^- |\widehat{E}_k^-|, \tag{7.13}$$

$$-1 \preceq \mathbf{y}_k \preceq 1.$$

The optimization problem in (7.13) can be efficiently solved by quadratic programming. Note that the objective function in (7.13) is not convex and the global minimum cannot be reached. Suppose that the optimal solution of (7.13) is $\mathbf{y}_k^*$, we can update $\mathbf{y}_k$ as:

$$y_{ki} = \begin{cases} +1 & y_{ki}^* \geq 0 \\ -1 & y_{ki}^* < 0 \end{cases} \tag{7.14}$$

After that, the outputs of the objective function in (7.13) when using the newest labels $\mathbf{y}_k$ and the old labels $\mathbf{y}_k^{prev}$ are further compared to ensure the objective in (7.11) decreases and the constraints in (7.13) are satisfied. That is, all the exemplar labels in $\mathbf{y}_k$ will be further replaced by their previous labels in $\mathbf{y}_k^{prev}$ when any of the constraints in (7.11) is violated or:

$$(1 + \mathbf{y}_k)^{\mathrm{T}} \mathbf{M}_k (1 - \mathbf{y}_k) \geq (1 + \mathbf{y}_k^{prev})^{\mathrm{T}} \mathbf{M}_k (1 - \mathbf{y}_k^{prev}). \tag{7.15}$$

This step is necessary to guarantee the convergence of the learning algorithm since it can ensure the objective value in (7.11) always decreases.

Optimize w: Given exemplar labels $\{y_{ki}\}$, we can generate a new set of target-distractor pairs V. Then, we solve the ordinal regression problem to obtain the optimal $\mathbf{w}$:

$$\min_{\mathbf{w}, \{\xi_v\}} \frac{1}{2} \|\mathbf{w}\|_2^2 + C \sum_{v \in \mathcal{V}} \xi_v,$$

$$s.t. \ \ \forall v \in \mathcal{V}, \ \mathbf{w}^{\mathrm{T}} (\mathbf{x}_{ki} - \mathbf{x}_{kj}) \geq 1 - \xi_v, \xi_v \geq 0. \tag{7.16}$$

We can see that the formulation in (7.16) is similar to the traditional SVM, which is also a convex quadratic programming problem that aims to find a linear function to minimize the swapped instance orders (Joachims, 2006). Here, we use the Liblinear package (Fan et al, 2008) to solve this problem in linear time and also ensure that the objective value decreases.

The detailed learning process is listed in Algorithm 1. By iteratively performing these two steps, the algorithm will gradually converge since the objective value always decreases. In this work, we terminate the learning algorithm if a predefined number of iterations is reached or less than 0.1% of the exemplars change their labels in an iteration. Actually, the learning algorithm often converges within less than ten iterations, which is illustrated in the experiment Section.

Algorithm 2. The multi-instance learning to rank algorithm

Input   : Local visual attributes $\{\mathbf{x}_{ki}\}$, user labels $\{g_{ki}\}$, iteration numbers T.
Output: Model parameter $\mathbf{w}$, instance labels $\{y_{ki}\}$.
begin
   $t = 0$;
   $y_{ki}^0 = g_{ki}, \forall k, i$;
   solve the problem in (7.16) to obtain $\mathbf{w}^0$;
   foreach k **do**
      $(\mathbf{M}_k^0)_{ij} = \max(0, 1 - (\mathbf{x}_{ki} - \mathbf{x}_{kj})^{\mathrm{T}} \mathbf{w}^0)/4, \forall i, j$;
      $\epsilon_k^0 = (1 + \mathbf{y}_k^0)^{\mathrm{T}} \mathbf{M}_k^0 (1 - \mathbf{y}_k^0)$;
   end
   repeat
      $t \leftarrow t + 1$;
      foreach k **do**
         solve the problem in (7.13) to obtain $\mathbf{y}_k^*$;
         $\forall i, y_{ki}^t = \begin{cases} +1 & y_{ki}^* \geq 0 \\ -1 & y_{ki}^* < 0 \end{cases} \quad \epsilon_k^t = (1 + \mathbf{y}_k^t)^{\mathrm{T}} \mathbf{M}_k^{t-1} (1 - \mathbf{y}_k^t)$;
         if $\epsilon_k^t \geq \epsilon_k^{t-1}$
            or $(1 + \mathbf{g}_k)^{\mathrm{T}} (1 + \mathbf{y}_k^t) < 4\rho_k^+ |\widehat{E}_k^+|$
            or $(1 - \mathbf{g}_k)^{\mathrm{T}} (1 - \mathbf{y}_k^t) < 4\rho_k^- |\widehat{E}_k^-|$
         then $y_{ki}^t = y_{ki}^{t-1}, \forall i$;
      end
      solve the problem in (7.16) to obtain $\mathbf{w}^t$;
      foreach k, i, j **do**
         $(\mathbf{M}_k^t)_{ij} = \max(0, 1 - (\mathbf{x}_{ki} - \mathbf{x}_{kj})^{\mathrm{T}} \mathbf{w}^t)/4$;
      end
      $N_{var}^t = \frac{1}{2} \sum_{k,i} |y_{ki}^t - y_{ki}^{t-1}|$;
   until $t \geq T$ *or* $N_{var}^t < 0.001 \sum_k l_k$;
end

7.3.2.4 Visual Saliency Estimation

After learning the optimal ranking function from the training data, we use it to estimate the saliency map for a testing image. In this process, the ranking function outputs a real value $\phi(\mathbf{x}_{ki})$ for each patch e_{ki} in the testing image with respect to its local visual attributes $\mathbf{x}_{ki}$. Note that here the numerical value of $\phi(\mathbf{x}_{ki})$ is immaterial and this value is only used to assign each patch an integer rank $c_{ki} \in \{1, \ldots, N_k\}$. Among these ranks, the most salient patches will have smaller ranks and distractors will be associated with larger ones. In particular, two patches $\mathbf{x}_{ki}$ and $\mathbf{x}_{kj}$ will be assigned the same integer rank once $|\phi(\mathbf{x}_{ki}) - \phi(\mathbf{x}_{kj})|$ is smaller than a predefined threshold (e.g., 0.01). With this solution, we can effectively cope with noise or small model disturbance.

To reveal the distribution of salient patches in an image, these integer ranks should be further mapped into real saliency values. In the conversion, patches with small ranks should be popped-out while patches with large ranks need to

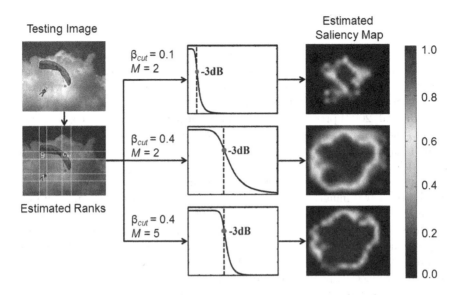

Fig. 7.6 Turning integer ranks to real values with Butterworth filter. Copyright © IEEE. All rights reserved. Reprinted, with permission, from (Li et al, 2012).

be suppressed. Meanwhile, the relative saliency correlations between any two patches should be preserved. Toward this end, a monotonically decreasing function with the output range [0,1] is required. Inspired by the response function of Butterworth low-pass filter which is smooth in the pass-band and decrease to zero in the stop band, we adopt the following function to map the integer rank $c_{ki} \in \{1, \dots, N_k\}$ into a real saliency value $s_{ki} \in [0, 1]$:

$$s_{ki} = \frac{1}{\sqrt{1 + (\frac{c_{ki}}{\beta_{cut} N_k})^{2M}}}, \qquad (7.17)$$

where $\beta_{cut} > 0$ is the normalized cut-off frequency (i.e., the -3dB point) and M is the order of Butterworth filter. As shown in Fig. 7.6, a larger β_{cut} can pop-out more image patches since the function starts to decrease at a larger rank. Meanwhile, the filter with a higher order leads to a steeper decrease in saliency values when the estimated ranks become higher than $\beta_{cut} N_k$. In this study, we empirically set $\beta_{cut} = 0.1$ and $M = 4$, which can successfully pop-out the entire salient objects and suppress most distractors (with zero saliency).

7.3.3 Experiments

In this Section, we evaluate the performance of the proposed approach on two public image datasets. The first dataset, denoted as **MSRA** (Liu et al, 2007), is a **regional saliency dataset** that consists of 5,000 images with obvious salient objects and "clean" background regions. For each image, Liu *et al.* (Liu et al, 2007) provided several highly consistent rectangles covering the entire salient object. Moreover, Achanta *et al.* (Achanta et al, 2009) selected 1,000 images from **MSRA** and manually generated the accurate contour for each salient object. In the experiments, we only use the 1,000 images with accurate object masks for training and testing. From these images, we randomly select 900 images with labeled rectangles to form the training subset, while the testing subset consists of the rest 100 images with accurate object masks. We observe that the rectangles provided by different subjects are highly consistent and the inter-subject ROC can reach as high as 0.97. In the training subset, an image patch will be treated as a positive training sample once it is labeled as positive by any subject.

The second dataset **MIT** is an **eye-fixation dataset** provided by Judd *et al.* (Judd et al, 2009). This dataset consists of 1,003 images with rich contents. For each image, a fixation density map is generated with the eye-tracking devices to depict the distribution of salient objects. For this dataset, we select 903 images to construct the training subset and the rest 100 images are used as the testing subset. In the training subset, an image patch will be considered as a positive training sample once eye fixations from any subject fall on it.

On these two datasets, three experiments are conducted. In the first experiment, the proposed approach is compared with several baselines on **MSRA**. The main objective of this experiment is to demonstrate the effectiveness of various components of the proposed approach, such as the utilization of representative exemplars and training ranking model in the multi-instance learning framework. These baselines include:

- SVM-AP, which uses the *representative* exemplars generated by affinity propagation to train a typical SVM classification model.
- SVM-Random, which *randomly* selects the same numbers of positive/negative exemplars as SVM-AP and train a typical SVM classification model.
- Ranking-AP, which uses the same exemplars as SVM-AP to train a ranking model by directly solving (7.16).
- Ranking-Random, which uses the same exemplars as SVM-Random to train a ranking model by directly solving (7.16).

In the second and the third experiments, the multi-instance learning to rank approach, denoted as **MILR**, is compared with 11 state-of-the-art algorithms on **MSRA** and **MIT**, respectively. The main objectives of these two experiments are: 1) to demonstrate the effectiveness of **MILR** when the training data contain rich pseudo targets (**MSRA**) or pseudo distractors (**MIT**), and 2) to show the advantages and disadvantages of various approaches in visual saliency computation. In these two experiments, the eleven algorithms which are used for comparisons can be grouped into two categories, including:

1. **Bottom-up approaches**, include seven bottom-up models such as Itti98 (Itti et al, 1998), Hou07 (Hou and Zhang, 2007), Harel07 (Harel et al, 2007), Achanta09 (Achanta et al, 2009), Wang10 (Wang et al, 2010), Goferman10 (Goferman et al, 2010) and Cheng11 (Cheng et al, 2011). Note that in (Cheng et al, 2011) Cheng *et al.* proposed two algorithms for visual saliency computation. In this work, we only report the results from a regional contrast-based approach because it achieves better performance.
2. **Top-down approaches**, which can learn from the past visual stimuli and transfer the learned prior knowledge to the new stimuli. These approaches include Peters07 (Peters and Itti, 2007), Judd09 (Judd et al, 2009), Navalpakkam07 (Navalpakkam and Itti, 2007) and Freund03 (Freund et al, 2003). Note that Freund03 is a classical learning to rank approach. For fair comparison, we also use (7.17) to map its estimated ranks into real saliency values, in which we adopt the same parameters as in **MILR**.

In these three experiments, each algorithm was evaluated from multiple perspectives to demonstrate its advantages and disadvantages. Recall that the objective of visual saliency computation is to pop-out targets and suppress distractors, we need to test each algorithm from two perspectives:

1. **Ordering**: whether the targets are with smaller ranks when compared with the distractors (i.e., targets are more salient than distractors), and
2. **Amplitude**: whether the distractors are successfully suppressed (i.e., the saliency map should contain less "noise").

Toward this end, we adopt the following three evaluation criteria:

1. **Recall**, which corresponds to the probability that targets have smaller ranks than distractors, can be computed as:

$$\mathbf{Recall} = \frac{\sum_{k,i \neq j} [g_{ki} > g_{kj}]_{\mathbf{I}} [s_{ki} > s_{kj}]_{\mathbf{I}}}{\sum_{k,i \neq j} [g_{ki} > g_{kj}]_{\mathbf{I}}}, \qquad (7.18)$$

where $[\mathbf{x}]_{\mathbf{I}} = 1$ if event $\mathbf{x}$ holds and $[\mathbf{x}]_{\mathbf{I}} = 0$ otherwise. Note that here the user labels $\{g_{ki}\}$ can be generated either from the accurate object masks on **MSRA** or from the fixation density maps on **MIT**. From the definition in (7.18), we can see that **Recall** equals to 1 when all salient targets have smaller ranks than distractors.

2. **Precision**, which equals to the ratio of energy assigned to the salient targets, can be computed as:

$$\textbf{Precision} = \frac{\sum_k \sum_i [g_{ki} > 0]_{\mathbf{1}} s_{ki}}{\sum_k \sum_i s_{ki}}. \tag{7.19}$$

From the definition in (7.19), we can see that an algorithm which can pop-out only targets and suppress all distractors will obtain a **Precision** of 1. Actually, **Recall** only focuses on the *ordering* of the predictions, which is quite similar to the most prevalent evaluation metric AUC (Area Under the ROC Curve). However, such ordering characteristic is insufficient to fully capture the deviation between ground-truth saliency maps and estimated saliency maps. Toward this end, **Precision** is introduced to further evaluate an algorithm from the perspective of *amplitude*.

3. **F-Score**, which is the weighted average of **Recall** and **Precision**, can be computed as:

$$\textbf{F-Score} = \frac{(1 + \alpha) \times \textbf{Recall} \times \textbf{Precision}}{\alpha \times \textbf{Recall} + \textbf{Precision}}, \tag{7.20}$$

where $\alpha > 0$ is a parameter to balance **Recall** and **Precision**. In different scenarios, α can be changed to select the best model for visual saliency computation. For example, a large α emphasizes the precise detection of salient targets, while a small α requires the algorithm to pop-out the entire salient targets (even with probable distractors). In this study, we set $\alpha = 1.0$ to equally balance **Recall** and **Precision**.

7.3.3.1   Comparisons with Baselines

In this experiment, **MILR** is compared with four baselines on **MSRA** to demonstrate the effectiveness of various components. In the experiment, each model is trained using the 900 images along with the labeled rectangles, in which the training data contain rich pseudo targets. First, we evaluate the effectiveness of **MILR** in removing label ambiguities by using the accurate object masks of the 900 images in the training subset. Specifically, given the accurate object masks of 900 training images, we count the percentages of real targets, pseudo targets, real distractors and pseudo distractors before and after the learning process of **MILR**. From the results, we can find that the training images contain rich pseudo targets (17.0%) and few pseudo distractors (0.2%). After the learning process of **MILR**, most pseudo targets can be correctly relabeled as real distractors (*i.e.*, the percentage significantly reduces from 17.0% to 7.3%), while some of the real targets are also wrongly labeled (*i.e.*, the percentage also slightly reduces from 20.3% to 18.4%). The overall label ambiguities are now greatly reduced. Moreover, we can see from the results that the ratio of real targets in the positive bags increases from 54.4% to 71.6%. With these "cleaner" training instances, **MILR** can achieve better performance.

These models are then tested on the 100 images which have accurate masks for the salient objects. The results are shown in Table 7.2. From Table 7.2, we can see that SVM-AP outperforms SVM-Random and Ranking-AP is better than Ranking-Random in terms of **Recall**, **Precision** and **F-Score**. This is due to their different exemplar selection methods. Specifically, salient objects may comprise of multiple components with diverse appearance. In this case, affinity propagation can easily find the most representative exemplars for each component, while random sampling method may ignore the small components. Although the sampling rate can be increased to guarantee that at least one exemplar is selected from each component, this will lead to much higher computational cost. Moreover, with the random sampling method, it is more likely that many exemplars will be chosen from the large components when compared with small components, which may also lead to the over-fitting problem.

In Table 7.2, we can also observe that the ranking-based approaches, including Ranking-AP, Ranking-SVM and **MILR**, perform much better than the binary-classification based approaches (i.e., SVM-AP and SVM-Random). The explanation is that it is generally more difficult to directly classify whether an object is salient or not. When estimating the visual saliency value for an object, its visual attributes as well as the context information should be considered simultaneously. That is, an object will be classified as a salient one only if it wins the competition between all the candidate objects in the image. By considering such mutual competition in the learning process, the ranking-based approaches can perform much better.

MILR achieves the best performance when compared with all these four baselines. When compared with Ranking-AP, the relative **F-Score** improvement of **MILR** is 3.1%. The major difference between **MILR** and Ranking-AP is that Ranking-AP takes all the exemplars in the labeled rectangles as real targets, while **MILR** allows the existence of pseudo targets. By iteratively relabeling these pseudo targets as distractors, **MILR** can better utilize the correlations between real targets and real distractors. That also explains why **MILR** (**Precision**=0.757) is more accurate for locating the salient object when compared with Ranking-AP (**Precision**=0.722).

Table 7.2 Performance of **MILR** and baselines on **MSRA**. Copyright © IEEE. All rights reserved. Reprinted, with permission, from (Li et al, 2012).

Algorithm	Recall	Precision	F-Score
SVM-AP	0.822	0.501	0.623
SVM-Random	0.792	0.462	0.584
Ranking-AP	0.922	0.722	0.809
Ranking-Random	0.917	0.708	0.799
MILR	**0.929**	**0.757**	**0.834**

Fig. 7.7 The performance variations of **MILR** using different patch sizes. Copyright © IEEE. All rights reserved. Reprinted, with permission, from (Li et al, 2012).

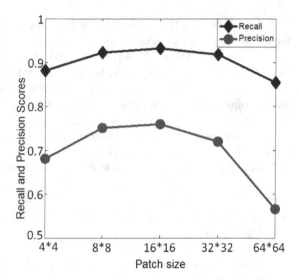

Moreover, we conduct an experiment to demonstrate the effectiveness of the location feature (i.e., the 43th feature), in which we remove the location feature in $\mathbf{x}_{ki}$ and report the results from Ranking-AP and **MILR** using the rest 42-dimensional features. We observe that the **F-Scores** of Ranking-AP and **MILR** are decreased to 0.787 and 0.810, respectively, which demonstrates the location feature is an important cue for visual saliency computation due to the center-bias property of human visual attention.

We also conduct an experiment to show the performance variations of **MILR** when using different patch sizes. As shown in Fig. 7.7, the 16×16 patches used in this study can achieve the highest **Recall** and **Precision**. In contrast, smaller patches (e.g., 4×4 and 8×8) are often too small to represent the local image characteristics and also lead to high computational complexity. Meanwhile, larger patches (e.g., 32×32 and 64×64) often fail to represent the image details and also may miss the small salient objects. It is still an open problem to determine the optimal patch sizes, which will be investigated in the future.

Finally, an important issue is the convergence speed of the learning algorithm. Although we have discussed that the output of the objective function in (7.9) is monotonically decreasing, the convergence speed is also very important and may be influenced by the predefined parameters (e.g., ρ_k^+ and ρ_k^-). Here we conduct the training process several times with $\rho_k^+ \in \{1/|\widehat{E}_k^+|, 20\%, 40\%, 60\%, 80\%\}$ (i.e., the predefined minimal percentage of real targets in each positive bag). As shown in Fig. 7.8, we find that **MILR** can converge within less than 10 iterations in all the cases and stringent constraints (e.g., $\rho_k^+ = 80\%$) may lead to faster convergence rate. We also have similar observations on **MIT** dataset. Therefore, we empirically set

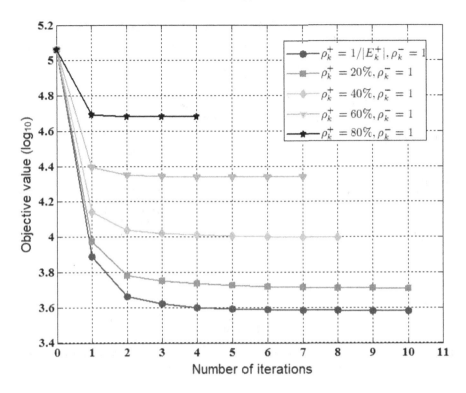

$\rho_k^+ = 60\%, \rho_k^- = 100\%$ for **MSRA** dataset and $\rho_k^+ = 95\%, \rho_k^- = 50\%$ for **MIT** dataset in all these three experiments.

7.3.3.2   Performance on Regional Saliency Dataset

In this experiment, **MILR** is compared with 11 state-of-the-art approaches on **MSRA** to demonstrate the advantages and disadvantages of all these approaches. On this dataset, all the learning-based approaches are trained using 900 images with labeled rectangles, which contain rich pseudo targets. After that, all these approaches are tested on the rest 100 images and the estimated saliency maps are compared with the binary masks of salient objects. The **Recall**, **Precision** and **F-Score** are reported in Table 7.3 and some representative results are illustrated in Fig. 7.11. Moreover, we also plot the ROC curves of these approaches in Fig. 7.9 and Fig. 7.10 to give a better view of their advantages and disadvantages.

From Table 7.3, we can see that **MILR** outperforms all the other methods in term of **F-Score**. From Table 7.3 and Figs. 7.9-7.11, we observe that these

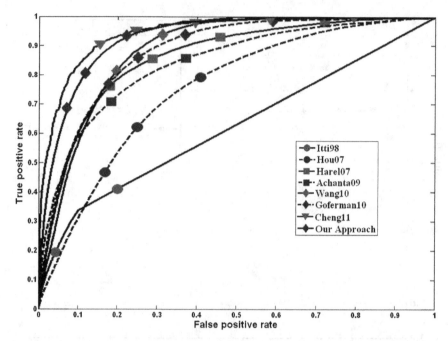

Fig. 7.9 ROC curves of **MILR** and bottom-up approaches on **MSRA**. Copyright © IEEE. All rights reserved. Reprinted, with permission, from (Li et al, 2012).

approaches can be grouped into three categories and their performances are tightly correlated with their definitions on visual saliency:

1) The first category contains all the seven bottom-up approaches and **visual saliency is defined as uniqueness or rareness**. The main **advantage** of these approaches is that salient objects are unique or rare in most cases, leading to relatively high recall rates. However, the main **disadvantage** is that it is often difficult to properly define such uniqueness or rareness. For example, some salient objects may be even larger than the distractors in certain images (e.g., the third row in Fig. 7.11) and their internal regions are no longer unique or rare. Moreover, a complex object may be unique or rare in an image, while some of its components can be also unique or rare when compared with the whole object (e.g., the last four rows in Fig. 7.11). In these cases, it is much more difficult to predefine the uniqueness or rareness to locate the entire salient object other than some of its components. Although Cheng11 can achieve the highest **Recall** by defining such uniqueness or rareness in object level, they also fail to remove the background objects from the estimated saliency maps, leading to low **Precision** and **F-Score**. From Fig. 7.9 and Fig. 7.10, we also observe Cheng11 slightly outperforms **MILR** in terms of the **ROC** curve. Similar as Recall, ROC curve only reflects the "ordering" of the predictions and it cannot well reveal the "precision" of an algorithm in locating salient objects. For instance, if

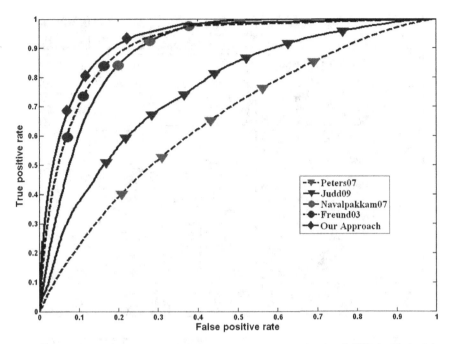

Fig. 7.10 ROC curves of **MILR** and top-down approaches on **MSRA**. Copyright © IEEE. All rights reserved. Reprinted, with permission, from (Li et al, 2012).

we iteratively convolve the estimated saliency maps with a Gaussian kernel, the **ROC** curve can be gradually improved, while the precision will generally decrease. To sum up, a loose definition of uniqueness or rareness will incorporate much noise in the estimated saliency maps, while a strict definition will fail to detect the entire saliency objects.

Table 7.3 Performance of Various Approaches on **MSRA**. Copyright © IEEE. All rights reserved. Reprinted, with permission, from (Li et al, 2012).

Algorithm		Recall	Precision	F-Score
Bottom-up	Itti98	0.333	0.582	0.424
	Hou07	0.754	0.372	0.498
	Harel07	0.853	0.395	0.540
	Achanta09	0.842	0.491	0.620
	Wang10	0.894	0.511	0.650
	Goferman10	0.878	0.453	0.598
	Cheng11	**0.961**	0.464	0.626
Top-down	Peters07	0.672	0.236	0.349
	Judd09	0.743	0.306	0.433
	Navalpakkam07	0.892	0.374	0.527
	Freund03	0.911	0.705	0.795
	MILR	0.929	**0.757**	**0.834**

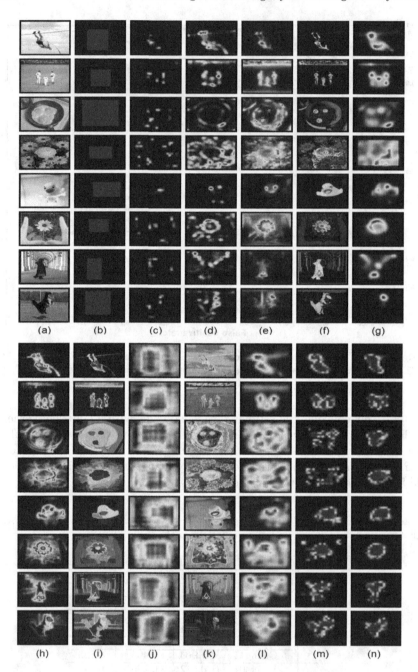

Fig. 7.11 Representative results of different approaches on **MSRA**. (a) Original images; (b) Binary masks of salient objects; and saliency maps estimated by (c) Itti98; (d) Hou07; (e) Harel07; (f) Achanta09; (g) Wang10; (h) Goferman10; (i) Cheng11; (j) Peters07; (k) Judd09; (l) Navalpakkam07; (m) Freund03; (n) **MILR**. Copyright © IEEE. All rights reserved. Reprinted, with permission, from (Li et al, 2012).

2) The second category contains three top-down approaches (Peters07, Judd09 and Navalpakkam07), in which **visual saliency is solely determined by local visual attributes**. This assumption, however, is often too strong for the general cases. As shown in Fig. 7.11(j)-(l), high saliency values may be assigned to distractors if these distractors used to be salient in the training data. Moreover, another main disadvantage of these approaches is that the label ambiguities are not considered. On the regional saliency dataset (e.g., **MSRA**), labeling a salient object with rectangles may generate many pseudo targets. If these pseudo targets are directly treated as real positives, the trained model may be biased and the estimated saliency maps will contain rich noise. That also explains that Peters07, Judd09 and Navalpakkam07 all achieve low **Precisions**.

3) The third category contains Freund03 and **MILR**, which demonstrate impressive performances in terms of **Recall** and **Precision**. The success of these two approaches lies in the assumption that **visual saliency is jointly determined by local visual attributes and spatial/spatiotemporal contexts**. The influences of this assumption are twofold: 1) the learning process can find the most effective visual features to distinguish real targets from real distractors; 2) by using these discriminate features in the testing stage, the saliency value of an image patch is determined by its "rank," while such "rank" is actually related to the local visual attributes of all candidate image patches. Moreover, **MILR** outperforms Freund03 by 4.9% in term of relative **F-Score**. This indicates that the multi-instance learning to rank framework can effectively remove the label ambiguities in the training data. One probable drawback of **MILR** is that $O(N)$ instances will generate $O(N^2)$ instance pairs for training. Fortunately, **MILR** can remove a great portion of redundancies from the training data by using the affinity propagation algorithm. By using a limited number of exemplars in a learning framework that rapidly converges, the computational cost thus becomes acceptable.

7.3.3.3   Performance on Eye-Fixation Dataset

In this experiment, **MILR** is compared with 11 state-of-the-art approaches on **MIT** to demonstrate the advantages and disadvantages of all these approaches. On this dataset, all the learning-based approaches are trained using 903 images with fixation-density maps (i.e., with rich pseudo distractors). After that, all these approaches are tested on the rest 100 images and the estimated saliency maps are compared with the fixation density map. The **Recall**, **Precision** and **F-Score** are reported in Table 7.4 and some representative results are illustrated in Fig. 7.14. Moreover, we also plot the ROC curves of these approaches in Fig. 7.12 and Fig. 7.13.

From Table 7.4, we can see that **MILR** also demonstrates the best overall performance (i.e., the highest **F-Score**) when the training data contain rich pseudo distractors. From Table 7.3 and Table 7.4, we can see that most

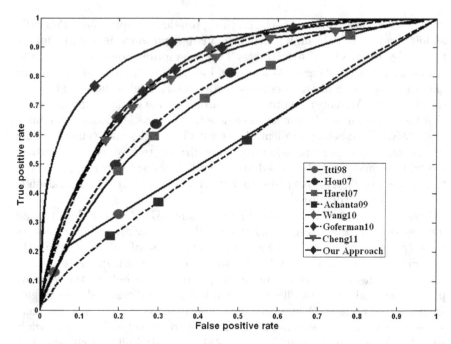

Fig. 7.12 ROC curves of **MILR** and bottom-up approaches on **MIT**. Copyright
© IEEE. All rights reserved. Reprinted, with permission, from (Li et al, 2012).

bottom-up approaches demonstrate comparable performance on these two
datasets. However, Cheng11 performs much worse on the eye-fixation dataset
when compared with the **MSRA** dataset. The main explanation is that
Cheng11 simply segments images into objects and then directly compute
the regional saliency in the object level. When the salient objects have relatively simple appearance and the background regions are relatively "clean"
(e.g., the images in **MSRA**), Cheng11 can easily segment and pop-out the
entire salient object from the background. However, the object segmentation
algorithm may fail when processing the images with rich contents (e.g., the
complex images in Fig. 7.14), making the **Recall** decrease remarkably.

By comparing the representative results in Fig. 7.11 and Fig. 7.14, we can
see that pseudo targets and pseudo distractors have different influences on
the learning-based approaches. As shown in Fig. 7.11(j)-(m), visual saliency
models directly trained with pseudo targets often introduce rich noise into
the estimated saliency maps. The explanation is that some distractors may
be wrongly classified as targets by these models. In contrast, many targets
will be wrongly classified as distractors due to the existence of pseudo distractors in the training data (as shown in Fig. 7.14(j)-(m)). Among all these
learning-based approaches, Judd09 tried to deal with this problem by randomly selecting training samples only from the most salient/nonsalient regions, leading to a relatively high recall rate. However, their **Precision** is far

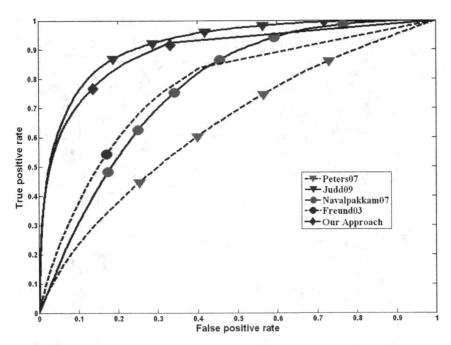

Fig. 7.13 ROC curves of **MILR** and top-down approaches on **MIT**. Copyright © IEEE. All rights reserved. Reprinted, with permission, from (Li et al, 2012).

from satisfactory since the randomly selected samples may not well represent the entire salient objects and backgrounds. Actually, it is more difficult to achieve a high **Precision** than a high **Recall** on **MIT** dataset since user fixations are often very sparse. In contrast, **MILR** starts with inaccurate instance labels and then iteratively update these labels to recover the actual

Table 7.4 Performance of Various Approaches on **MIT**. Copyright © IEEE. All rights reserved. Reprinted, with permission, from (Li et al, 2012).

Algorithm		Recall	Precision	F-Score
Bottom-up	Itti98	0.171	0.531	0.258
	Hou07	0.653	0.412	0.505
	Harel07	0.646	0.367	0.468
	Achanta09	0.514	0.322	0.396
	Wang10	0.763	0.502	0.606
	Goferman10	0.729	0.450	0.557
	Cheng11	0.706	0.401	0.511
Top-down	Peters07	0.596	0.310	0.407
	Judd09	**0.821**	0.406	0.543
	Navalpakkam07	0.686	0.392	0.499
	Freund03	0.616	0.528	0.569
	MILR	0.740	**0.770**	**0.755**

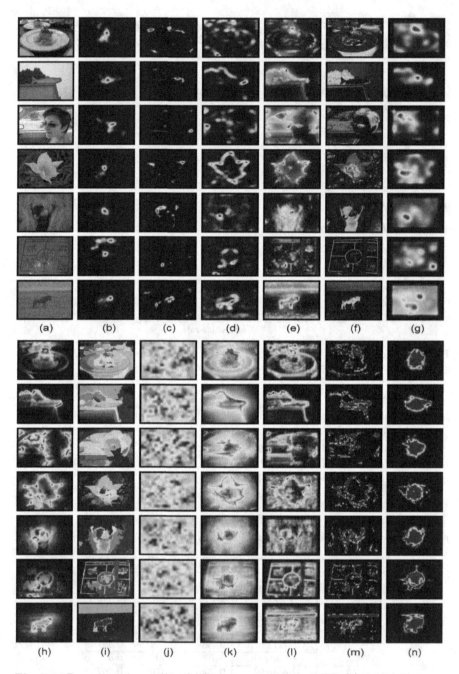

Fig. 7.14 Representative results of different approaches on **MIT**. (a) Original images; (b) Fixation density maps; and saliency maps estimated by (c) Itti98; (d) Hou07; (e) Harel07; (f) Achanta09; (g) Wang10; (h) Goferman10; (i) Cheng11; (j) Peters07; (k) Judd09; (l) Navalpakkam07; (m) Freund03; (n) **MILR**. Copyright © IEEE. All rights reserved. Reprinted, with permission, from (Li et al, 2012).

collection of targets and distractors. Therefore, the **Precision** of **MILR** is much higher than all the other approaches.

7.4 Notes

Observing that most user data are inaccurate or inadequate to directly train visual saliency models, we propose two approaches in this Chapter to cope with such label ambiguities. By embedding the sparse positive and unlabeled data into a cost-sensitive learning to rank framework, the ambiguities in video fixations can be well addressed. For ambiguities in image fixations or labeled rectangles, a multi-instance learning to rank framework is proposed to iteratively recover the correct instance labels and train the saliency model. The proposed learning algorithm can focus on the correlations between real targets and real distractors. Extensive experiments and comparisons demonstrate that these approaches achieve impressive improvement in the overall performance. Moreover, the estimated saliency maps are usually less "noisy" due to the successful removal of pseudo targets, and at the same time the entire salient object can pop-out due to the removal of pseudo distractors. We believe the saliency maps generated from these two approaches can facilitate further image-based applications such as object segmentation or content-based image retrieval.

References

Achanta, R., Hemami, S., Estrada, F., Süsstrunk, S.: Frequency-tuned salient region detection. In: Preceedings of the IEEE Conference on Computer Vision and Pattern Recognition (CVPR), pp. 1597–1604 (2009), doi:10.1109/CVPR.2009.5206596

Andrews, S., Tsochantaridis, I., Hofmann, T.: Support vector machines for multiple-instance learning. In: Advances in Neural Information Processing Systems (NIPS), pp. 561–568 (2003)

Bunescu, R.C., Mooney, R.J.: Multiple instance learning for sparse positive bags. In: Proceedings of the 24th International Conference on Machine Learning, Corvalis, OR, United states, vol. 227, pp. 105–112 (2007), doi:10.1145/1273496.1273510

Cheng, M.M., Zhang, G.X., Mitra, N., Huang, X., Hu, S.M.: Global contrast based salient region detection. In: Preceedings of the IEEE Conference on Computer Vision and Pattern Recognition (CVPR), pp. 409–416 (2011), doi:10.1109/CVPR.2011.5995344

Dempster, A., Laird, N., Rubin, D.: Maximum likelihood from incomplete data via the EM algorithm. Journal of the Royal Statistical Society, Series B 39(1), 1–38 (1977)

Fan, R.E., Chang, K.W., Hsieh, C.J., Wang, X.R., Lin, C.J.: Liblinear: A library for large linear classification. Journal of Machine Learning Research 9, 1871–1874 (2008)

Freund, Y., Iyer, R., Schapire, R.E., Singer, Y.: An efficient boosting algorithm for combining preferences. Journal of Maching Learning Research 4, 933–969 (2003)

Frey, B.J., Dueck, D.: Clustering by passing messages between data points. Science 315, 972–976 (2007)

Goferman, S., Zelnik-Manor, L., Tal, A.: Context-aware saliency detection. In: Proceedings of the IEEE Conference on Computer Vision and Pattern Recognition (CVPR), pp. 2376–2383 (2010), doi:10.1109/CVPR.2010.5539929

Guo, C., Ma, Q., Zhang, L.: Spatio-temporal saliency detection using phase spectrum of quaternion fourier transform. In: Preceedings of the IEEE Conference on Computer Vision and Pattern Recognition (CVPR), pp. 1–8 (2008), doi:10.1109/CVPR.2008.4587715

Harel, J., Koch, C., Perona, P.: Graph-based visual saliency. Advances in Neural Information Processing Systems (NIPS), 545–552 (2007)

Hou, X., Zhang, L.: Saliency detection: A spectral residual approach. In: Proceedings of the IEEE Conference on Computer Vision and Pattern Recognition (CVPR), pp. 1–8 (2007), doi:10.1109/CVPR.2007.383267

Itti, L.: Crcns data sharing: Eye movements during free-viewing of natural videos. In: Collaborative Research in Computational Neuroscience Annual Meeting, Los Angeles, California (2008)

Itti, L., Baldi, P.: A principled approach to detecting surprising events in video. In: Preceedings of the IEEE Conference on Computer Vision and Pattern Recognition (CVPR), vol. 1, pp. 631–637 (2005), doi:10.1109/CVPR.2005.40

Itti, L., Koch, C.: Computational modeling of visual attention. Nature Review Neuroscience 2(3), 194–203 (2001)

Itti, L., Koch, C., Niebur, E.: A model of saliency-based visual attention for rapid scene analysis. IEEE Transactions on Pattern Analysis and Machine Intelligence 20(11), 1254–1259 (1998), doi:10.1109/34.730558

Joachims, T.: Training linear svms in linear time. In: Proceedings of the 12th ACM SIGKDD International Conference on Knowledge Discovery and Data Mining (KDD 2006), pp. 217–226. ACM, New York (2006), doi:10.1145/1150402.1150429

Judd, T., Ehinger, K., Durand, F., Torralba, A.: Learning to predict where humans look. In: Proceedings of the IEEE International Conference on Computer Vision (ICCV), pp. 2106–2113 (2009), doi:10.1109/ICCV.2009.5459462

Kienzle, W., Wichmann, F.A., Scholkopf, B., Franz, M.O.: A nonparametric approach to bottom-up visual saliency. Advances in Neural Information Processing Systems (NIPS), 689–696 (2007)

Li, J., Tian, Y., Huang, T., Gao, W.: Cost-sensitive rank learning from positive and unlabeled data for visual saliency estimation. IEEE Signal Processing Letters 17(6), 591–594 (2010), doi:10.1109/LSP.2010.2048049

Li, J., Xu, D., Gao, W.: Removing label ambiguity in learning-based visual saliency estimation. IEEE Transactions on Image Processing 21(4), 1513–1525 (2012), doi:10.1109/TIP.2011.2179665

Liu, T., Sun, J., Zheng, N.N., Tang, X., Shum, H.Y.: Learning to detect a salient object. In: Proceedings of the IEEE Conference on Computer Vision and Pattern Recognition (CVPR), pp. 1–8 (2007), doi:10.1109/CVPR.2007.383047

Navalpakkam, V., Itti, L.: Search goal tunes visual features optimally. Neuron 53, 605–617 (2007)

Peters, R., Itti, L.: Beyond bottom-up: Incorporating task-dependent influences into a computational model of spatial attention. In: Proceedings of the IEEE Conference on Computer Vision and Pattern Recognition (CVPR), pp. 1–8 (2007), doi:10.1109/CVPR.2007.383337

Tseng, P.H., Carmi, R., Cameron, I.G.M., Munoz, D.P., Itti, L.: Quantifying center bias of observers in free viewing of dynamic natural scenes. Journal of Vision 9(7), 4, 1–16 (2009), doi:10.1167/9.7.4

Wang, W., Wang, Y., Huang, Q., Gao, W.: Measuring visual saliency by site entropy rate. In: Preceedings of the IEEE Conference on Computer Vision and Pattern Recognition (CVPR), pp. 2368–2375 (2010), doi:10.1109/CVPR.2010.5539927

Zhai, Y., Shah, M.: Visual attention detection in video sequences using spatiotemporal cues. In: Proceedings of the 14th Annual ACM International Conference on Multimedia (MULTIMEDIA 2006), pp. 815–824. ACM, New York (2006), doi:10.1145/1180639.1180824

Chapter 8
Saliency-Based Applications

This Chapter introduces several saliency-based applications in the fields of computer vision and multimedia analysis. We aim to demonstrate that by simulating the saliency mechanism in human vision system, computer can process visual information as human vision system does and the processing results can better meet human perception.

8.1 Overview

With the rapid development of Internet, the amount of images and videos is now growing explosively. Based on these massive images and videos, it is necessary to develop intelligent computer vision and multimedia applications to meet various user requirements.

Generally, an application can be called intelligent if it provides service that is tightly correlated with the important image or video content while meeting the interest of various users. From this definition, we can see that visual saliency plays an important role for such intelligent computer vision and multimedia applications. First, from the definition of visual saliency, salient subsets correspond to important image or video content. By focusing on these salient subsets, the processing results can be tightly correlated with the major content of image and video. Second, interesting targets are visually salient (Elazary and Itti, 2008). Thus visual saliency can reflect the user interest to some extent. By emphasizing the salient content, the application may provide user-targeted services according to their interests.

In the rest part of this Chapter, we will introduce several saliency-based applications in six categories, including retargeting, advertising, retrieval, summarization, compression and recognition. For each category, one or two representative approaches will be introduced.

8.2   Retargeting

Today, images or videos are often watched over various display devices with different resolutions. On these devices, adding black bars and uniform scaling are two traditional methods to display videos with different aspect ratios and resolutions. However, adding black bars cannot make full use of the whole screen, while uniform scaling may distort important video content. Therefore, it is necessary to develop adaptive image/video displaying techniques such as content-aware retargeting to enhance the viewing experience while preserving appearances of salient objects. Here we will first introduce the image retargeting technique, followed by the video retargeting approach.

8.2.1   Image Retargeting

To address the problem of content-aware image retargeting, a classic approach, called seam carving, was proposed in (Avidan and Shamir, 2007). A seam is an optimal 8-connected path of pixels on a single image from top to bottom, or left to right, where optimality is defined by an image energy function. By carving out or inserting seams in horizontal or vertical directions, images can be resized to any resolution or aspect ratio.

We can see that the key issue for seam carving is to find an image energy function for seam detection. Actually, the image energy can be defined by various visual saliency measures as well as user interactions. In this process, the salient map defines the important visual content and prevents it from being removed. For an image $\mathbf{I}$, let $e(\mathbf{I})$ be its saliency map, the question is how to chose the pixels to be removed?

Intuitively, the goal is to remove unnoticeable pixels that blend with their surroundings, and such pixels often appear in less salient regions. Without loss of generality, we assume the goal is to find the optimal vertical seam from $e(\mathbf{I})$ to reduce the image width. To achieve this goal while keeping pixels with high saliency values, the following strategies may help:

1. Remove the pixels with lowest energy in ascending order. But it destroys the rectangular shape of the image, because a different number of pixels may be removed from each row.
2. Remove an equal number of low energy pixels from every row. This preserves the rectangular shape of the image but destroys the image content by creating a zigzag effect.
3. Crop a sub-window with the width of the target image that contains the highest energy, or remove whole columns with the lowest energy. But artifacts might appear in the resulting image.

To overcome the drawbacks of these strategies, seam carving is proposed, which will be less restrictive than cropping or column removal, but can

preserve the image content better than single pixel removal. Formally, let
I be an $n \times m$ image and a vertical seam can be defined as:

$$\mathbf{s}^{\mathbf{x}} = \{(x(i), i)\}_{i=1}^{n}, \quad \forall i, \ |x(i) - x(i-1)| \leq 1, \tag{8.1}$$

where x is a mapping from $[1, \ldots, n]$ to $[1, \ldots, m]$ (i.e., the column index
in the ith row). That is, a vertical seam is an 8-connected path of pixels in
the image from top to bottom, containing only one pixel in each row of the
image. Similarly, a horizontal seam can be defined as:

$$\mathbf{s}^{\mathbf{y}} = \{(j, y(j))\}_{j=1}^{m}, \quad \forall j, \ |y(j) - y(j-1)| \leq 1. \tag{8.2}$$

Note that removing the pixels of a seam from an image has only a local
effect: all the pixels of the image are shifted left (or up) to compensate for
the missing path. The visual impact is noticeable only along the path of the
seam, leaving the rest of the image intact. Note also that one can replace the
constraint $|x(i) - x(i-1)| \leq 1$ with $|x(i) - x(i-1)| \leq k$, and get either a simple
column for k = 0, a piecewise connected or even completely disconnected set
of pixels for any value $1 \leq k \leq m$.

Given the saliency map $e(\mathbf{I})$, the cost of a seam $\mathbf{s}^{\mathbf{x}}$ can be computed as
$E(\mathbf{s}) = \sum_{i=1}^{n} e(\mathbf{I}(x(i), i))$. Correspondingly, the optimal seam $\mathbf{s}^{*}$ is the one
that minimize the cost $E(\mathbf{s})$, which can be found using dynamic programming.
The first step is to traverse the image from the second row to the last row
and compute the cumulative minimum energy M for all possible connected
seams for each entry (i, j):

$$M(i, j) = e(i, j) + \min(M(i-1, j-1), M(i-1, j), M(i-1, j+1)), \tag{8.3}$$

At the end of this process, the minimum value of the last row in M will
indicate the end of the minimal connected vertical seam. Hence, the path of
the optimal seam can be backtracked from the minimum entry on M. The
definition of M for horizontal seams is similar. The overall framework of the
seam carving approach is summarized in Fig. 8.1. We can see that the seams
are mostly selected from background pixels, while the shape of foreground
object can be well preserved. For more details of conducting the carving,
please refer to (Avidan and Shamir, 2007).

8.2.2   Video Retargeting

Compared with image retargeting, video retargeting is much more challenging
since it should consider the temporal consistency of the spatially retargeted
results. Moreover, the salient content in each frame is jointly determined
by the spatial and temporal factors, making it necessary to process multiple
consecutive video frames simultaneously. Therefore, video retargeting usually

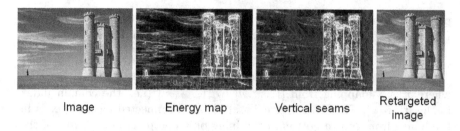

| Image | Energy map | Vertical seams | Retargeted image |

Fig. 8.1 Images are carved by maintaining the most conspicuous content

adopts an optimization framework and the computational complexity is much higher than image retargeting.

Here we will briefly introduce the video retargeting approach proposed in (Li et al, 2010). In this approach, the local spatiotemporal saliency is employed to measure the pixel-wise importance of video and the goal of video retargeting is to chop the less salient regions and keep the most informative parts. More formally, video retargeting can be formulated as the problem of optimizing a smooth trajectory for a cropping window, which can move along the spatial and temporal dimensions throughout the video (as shown in Fig. 8.2), to capture the most salient video content.

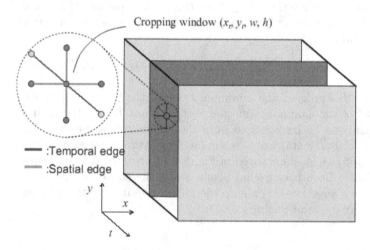

Fig. 8.2 Searching the optimal cropping window in the video volume. © (2010) Association for Computing Machinery, Inc. Reprinted by permission from (Li et al, 2010).

Toward this end, a trajectory-based video retargeting approach is proposed to gain the viewing experience of smooth and low distortion. In this approach, the problem of video retargeting is formulated as the process of finding out an optimal trajectory for a cropping window of a proper size to walk through the spatial and temporal volume (ST-volume). Along the trajectory, the moving cropping window is expected to capture the most saliency out of the ST-volume. This procedure is something like the gaze shifting process, where viewer always focuses on the most interesting parts along the video. In this approach, seeking the best one amongst all the trajectories of multi-scale cropping windows is able to model where to crop and how large ratio of areas to scale on the target resolution.

The system framework of the proposed approach is shown in Fig. 8.3. First, different types of local saliency features are extracted to give a final 2D-map for each video frame, producing a ST-volume for each video shot. Second, the ST-volume is separated into x channel and y channel by horizontal and vertical projection to compress the searching space. Third, for each channel, a weighed 2D graph is built on their saliency maps for a fixed size of cropping window, with the potential position of the cropping windows as its vertices, and edges measuring the cost of spatial and temporal movement. Then the problem of finding the optimal trajectory of the given size of cropping window can be formulated as a Max Flow/Min-Cut problem on x-axis and y-axis channel graphs. The minimum cuts correspond to the optimal x and y axis trajectories of the cropping window. The cropping window with the maximal scaled saliency and its optimized trajectory is picked up by ranking over all potential sizes of cropping windows. This procedure can be repeated to capture the remaining saliency by wiping out the saliency of the optimal trajectory from original ST-saliency volume. Finally, a merging procedure is used to improve the captured saliency.

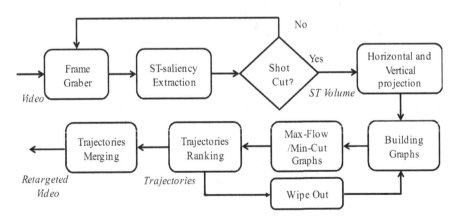

Fig. 8.3 Flowchart of the video retargeting approach in (Li et al, 2010). © (2010) Association for Computing Machinery, Inc. Reprinted by permission from (Li et al, 2010).

Note that the x and y variables of the trajectory are optimized separately in the proposed approach. Here we only demonstrate how to optimize the x

variables across a video shot of T frames. The estimation of the y variables can be solved similarly. Let $\mathbf{G}=(\mathbf{V},\mathbf{E})$ be a graph. The vertex set is denoted as $\mathbf{V}$ and the set of edges between vertexes as $\mathbf{E}$. Let each cropping window be represented by a vertex $v \in \mathbf{V}$ and horizontal saliency importance map of each frame can be sampled by m 1D cropping windows of width w. There are totally mT cropping windows in the graph of x-axis channel of the shot. An additional "source" node connects to the x_0 nodes at all frames, and a "sink" node connects to x_{max} nodes in all frames. As shown in Fig. 8.4, each vertex is four-connected to its neighbors. In particular, each vertex of the candidate cropping window is connected with its neighbors by spatial edge in x-axis and temporal edge in t-axis.

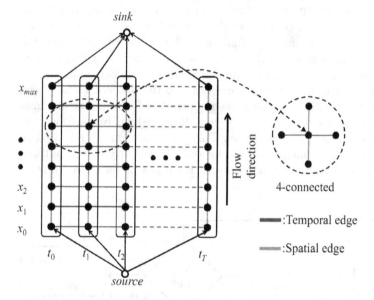

Fig. 8.4 Graph-based trajectory optimization in x-axis. © (2010) Association for Computing Machinery, Inc. Reprinted by permission from (Li et al, 2010).

On the graph, the nodes correspond to cropping windows and the edges are weighted by the cost of moving between cropping windows. The cost of an edge is derived from the average saliency of two vertices that it links. A cropping window is likely to move between two nodes with low transferring cost (i.e., with similar saliency), but the movement between targets and backgrounds are prevented. Since dynamic programming is not suitable in this case, the Max-Flow/Min-Cut is used to find the min-cut for the graph. Once the maximum flow is found, a min-cut the graph minimizes the sum of edge cost separates the source and sink. This cut also provides the trajectory with the minimal cost for a cropping window to go through the whole video shot. (i.e., a set of (x,y) values at $t_0,\ldots,t_T$).

Some representative retargeting results when using the proposed approach are given in Fig. 8.5. We can see that the retargeted frames can keep the most salient parts in each image when using different target resolutions. Subjective studies report that the viewing experience of the retargeted videos is satisfactory in nearly 80% of about the 1,500 testing video shots. Moreover, this approach is fully automatic, globally optimized, without any constraints on the movement of the cropping window. Despite the stage of shot detection, this approach can be implemented efficiently for real time retargeting.

Fig. 8.5 Representative retargeted video frames. 1st row: original frame; 2nd row: retargeted to 160 × 120 (red); 3rd row: retargeted to 120 × 120 (blue). © (2010) Association for Computing Machinery, Inc. Reprinted by permission from (Li et al, 2010).

8.3   Advertising

In recent years, the amount of videos on the Internet has increased tremendously. Thus it would be encouraging to provide incentive services such as personalized video recommendation and online video content association. In particular, video content association refers to the service that associates additional materials (e.g., texts, images and video clips) with the video contents to enrich the viewing experience and a promising application is content-aware video advertising. This new model of advertising is less intrusive, only displaying content-related ads in the user-targeted way. Therefore, content-aware video advertising presents a significant opportunity for marketers to extend the reach of their campaigns with compelling content (Gao et al, 2010).

There have been many approaches to associate ads such as logos and video clips with video contents. In these approaches, different advertising guidelines have been proposed. For example, some researches (Guo et al, 2009; Li et al, 2005; Chang et al, 2008) pointed out that a good advertising should not interrupt the viewing experience, while others argued that the ads should be relevant to video content (Mei et al, 2007; Wang et al, 2008) or user preference (Lekakos et al, 2001; Thawani et al, 2004) so as to enrich the viewing experience. To sum up, a smart advertising service should have the following characteristics:

1. **Non-intrusive**. The advertisement should not interrupt, clutter or delay the viewing experience;
2. **Content-related**. The advertisement should be relevant to the important video content; and
3. **User-targeted**. The advertisement should match individual preferences of different users.

Since it is often difficult to model individual preference, existing studies focus on the first two characteristics by solving one key problem: **when** to insert **what** to **where**. To solve this problem, visual saliency plays an important role since saliency is tightly correlated with attention. Therefore, inserting ads at non-attractive time points can be less intrusive. Moreover, salient subsets are visually important for content-aware advertising. Furthermore, non-salient regions can be good candidates for ad overlay. To sum up, visual saliency is inevitable to determine when, where and what to advertise.

In existing approaches, there are mainly two ways to associate ads with video contents, including in-banner and in-stream advertising. As shown in Fig. 8.6(a), in-banner advertising places ads on banners around the video window, while in-stream advertising refers to inserting or overlaying multimedia materials into video streams. Since the main difference between in-banner and in-stream advertising approaches is the placement of ad, we will mainly discuss in-stream advertising in this book.

In general, approaches for in-stream advertising can be grouped into two categories, including linear and non-linear advertising. In linear approaches, ad clips are inserted before, in the middle of or after the related video segment (as shown in Fig. 8.6(b)). Thus these approaches focus on the "when" and "what" problems by inserting content-related video clips to less-intrusive time points. At the insertion points, users are forced to watch ads which take over the full view of the video. Typically, linear advertising is prevalent but inevitably prolongs the original video and interrupts the viewing experience.

Another way for in-stream advertising is non-linear overlay, which provides ads in parallel to the video content. As shown in Fig. 8.6(c), the simplest way for non-linear advertising is to embed ad logos onto the video content. Different from the linear approaches, non-linear advertising focuses on the "what" and "where" problems. That is, overlaying contextually relevant image ads onto non-intrusive video positions. Ads inserted by these approaches can easily capture human visual attention, but the original video content is revised, making the ads more intrusive (McCoy et al, 2007a,b; Li et al, 2002). In the following parts, we will introduce approaches in these two categories.

(a) In-banner   (b) Linear in-stream   (c) Non-linear in-stream

Fig. 8.6 Three feasible advertising solutions

8.3.1 Linear Advertising

As a representative linear approach for in-stream advertising, Mei et al (2007) proposed to insert content-related ad clips into less intrusive temporal positions. The system framework of this approach is shown in Fig. 8.7. From Fig. 8.7, we can see that the advertising system consists of three main modules, including:

1. **Candidate ad selection**: selecting candidate ad clips from a large ad database (the top pathway of Fig. 8.7). In this module, ads in the database are ranked according to the textual relevance between ads and original video by using the query keywords (if a source video is reached by searching through a query), title, tags (a textual description provided by content providers or advertisers), and closed captions (if available). Such textual information is then used in a probabilistic model to rank the ads in the database.
2. **Candidate insertion point selection**: extracting visual-aural features from the source video to detect candidate time positions for ad insertion (the bottom pathway of Fig. 8.7). In this module, the original video is first divided into shots, each of which is then represented by a key-frame. After that, the discontinuity is computed to measure the content dissimilarity between two consecutive shots, while the attractiveness is measured according to the attention curve in (Ma et al, 2005). Given the attention values for each shot, the content attractiveness of an insertion point is highly related to the neighboring shots on both sides. Therefore, the attractiveness of an insertion point can be computed by weighted averaging the attention values of its neighboring shots.
   Given the discontinuity and attractiveness, insertion points can be thus selected as those time positions with high discontinuity and low attractiveness. One way for detecting ad insertion points can be finding peaks at the combined curve with discontinuity minus attractiveness. Moreover, the detection of ad insertion point should also consider the temporal

distribution of these points, as well as the global and local relevance between source video content and ad content.

3. **Optimization-based ad insertion**: matching ads and insertion points
   through the optimization algorithm (the middle pathway of Fig. 8.7).
   Given the discontinuity and attractiveness, this module optimizes the ad
   insertion by the local visual-aural relevance between each ad and the neighboring source video content on both sides of corresponding insertion point.
   The optimization process aims at selecting a subset of candidate insertion
   points and ads to maximize contextual relevance and minimize intrusiveness. For more details of the optimization process, please refer to (Mei
   et al, 2007).

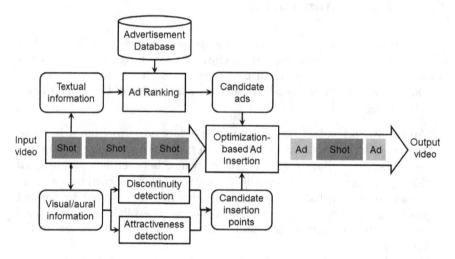

Fig. 8.7 Flowchart of the advertising approach in (Mei et al, 2007)

8.3.2   Non-linear Advertising

Non-linear advertising approaches focus on directly overlaying ads onto video
content. In these approaches, the role of visual saliency/attention is to determine what ads should be adopted and where to overlay these ads. Intuitively,
an efficient way to unobtrusively attract people's attention on ad is to emplace
it on those salient regions where people concentrate on watching. Toward this
end, Chang et al (2008) proposed to embed ad logos onto tennis video by attention calculation. By analyzing subject's attention when watching tennis
video, they concluded that court view frames could represent viewers' main
focus of the game proceeding in the temporal domain, while the locations of
players and balls were the attractive areas in the spatial domain. Therefore,
they proposed to emplace ad logos onto attractive regions to let audiences
see the advertisement consciously or unconsciously without interruption.

The system framework of this approach is shown in Fig. 8.8. In this system, court view shots in tennis video are detected first and camera calibration is then conducted to help locating the player in each frame. Based on the player's location, the attention region is detected for placing ads. In this manner, the position movement of the player gracefully spotlights the advertisement. Audiences would unconsciously reinforce their memory of the advertisement without missing the details of the game proceeding. Moreover, the size, transparency, and color distribution of the projected ad is harmonized according to the target court view frames to further limit the intrusiveness introduced. Subjective experiments reported that this system performed impressive in directing user's attention to advertisements during their free time.

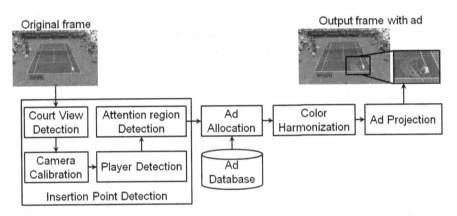

Fig. 8.8 Flowchart of the advertising approach in (Chang et al, 2008)

8.4   Retrieval

Since visual saliency can help to locate the important content in image and video, some studies adopt the saliency cues to improve the retrieval performance. The main idea is that visual features from salient objects can better represent the image or video than those from background regions. Here we will discuss how image retrieval can benefit from the usage of visual attention/saliency models.

In (Fu et al, 2006), an attention-driven image retrieval system is proposed. In this approach, image is interpreted as containing several perceptually salient objects as well as a background, where each object has a saliency value. The saliency values of attentive objects are then mapped to importance factors so as to facilitate the image retrieval. In the retrieval, images are matched when emphasizing the salient objects.

Without loss of generality, an image I is interpreted as T salient objects $\{O_i\}_{i=1}^T$ and the background O_{T+1}. Let $F_i \in [0,1]$ be the saliency value for O_i and $F_i > F_{i+1}$, the importance factors $\{f_i\}_{i=1}^{T+1}$ of these objects in the retrieval can be progressive computed using the following pseudocode:

- Step 1: Let $i = 1$, $\alpha = 1$;
- Step 2: $f_i = \frac{\alpha}{2}F_i + \frac{\alpha}{2}$
- Step 3: $\alpha = \alpha - f_i$;
- Step 4: $i = i + 1$;
- Step 5: If $i = T + 1$, $f_i = \alpha$, stop; otherwise goto Step 2.

In this manner, each object (and the background) in image I is assigned an importance value, and $\sum_{i=1}^{T+1} f_i = 1$. Note that the salient object will be emphasized with large importance value, despite its size.

Suppose that the query image $I_q = \{O_i\}_{i=1}^q$ and candidate image $I_c = \{O'_i\}_{i=1}^c$ has importance factors $\{f_i\}_{i=1}^q$ and $\{f'_i\}_{i=1}^c$, respectively. The distance matrix among objects in these two images is given by:

$$D = \{d_{i,j} = sd(O_i, O'_j), i = 1, \ldots, q; j = 1, \ldots, c\}, \qquad (8.4)$$

where $sd(O_i, O'_j)$ is the Euclidean distance between the feature vectors of object O_i and O'_j (e.g., HSV color histogram and texture histogram). Moreover, a weighting matrix, which indicates the importance of the element of the distance matrix, is defined as:

$$W = \{w_{i,j} = f_i \times f'_j, i = 1, \ldots, q; j = 1, \ldots, c\}. \qquad (8.5)$$

Moreover, a matching matrix is used to indicate the pair-wise matching between an object in the query image and the most similar object in the candidate image:

$$M = \{m_{i,j}, i = 1, \ldots, q; j = 1, \ldots, c\}, \qquad (8.6)$$

where

$$m_{i,j} = \begin{cases} 1, & \text{if } j = j_{min} \text{ and } j_{min} = \arg\min_j(d_{i,j}) \\ 0, & otherwise \end{cases}. \qquad (8.7)$$

For each row of the M matrix, only one element is 1 and all the other elements are 0. The element being 1 indicates the corresponding $d_{i,j}$ achieves the minimum difference in the row. Given the distance matrix, weighting matrix and matching matrix, the distance between two images is defined as:

$$(I_q, I_c) = \frac{\sum_{i=1}^q \sum_{j=1}^c m_{i,j} w_{i,j} d_{i,j}}{\sum_{i=1}^q \sum_{j=1}^c m_{i,j} w_{i,j}}, \qquad (8.8)$$

where a smaller distance indicates that two images have more similar objects and/or the matched objects are more important. As reported in the experiments of (Fu et al, 2006), such object-based matching can achieve impressive results.

Similar approaches can also be used for video retrieval. For example, Li and Lee (2007) first extracted a set of keyframes from each shot according to the motion and content complexity. The spatiotemporal saliency algorithm was applied on each key frame to extract a saliency map and a number of attentive regions. Consequently, the matching between video shots can be conducted by measuring the similarities between keyframes. Moreover, the similarity between keyframes can be computed according to the similarity between attentive regions as in (Fu et al, 2006).

8.5   Summarization

As an important application of visual attention/saliency model, a video summarization system is developed in (Ma et al, 2005) to demonstrate the effectiveness, robustness and generality of the attention framework. In this approach, a set of attention models (as shown in Fig. 8.9) are first generated by using various features such as motion, appearance, object and speech. These attention models include:

1. **Motion attention model.** For each video frame, a motion attention map is derived from motion vector field by jointly measuring the motion intensity, spatial coherence and temporal coherence. After that, a continuous motion attention curve along time axis is generated by averaging the motion attention maps on all frames.
2. **Static attention model.** For each frame, a center-surround spatial contrast map is first computed. From this map, a fuzzy growing algorithm is used to extract attractive regions. Finally, a static attention curve is formed by jointly considering the attended regions and the center-bias effect on each video frame.
3. **Face attention model.** The appearance of dominant faces in video usually attracts viewers' attention. Therefore, a face attention curve is generated according to the sizes and positions of all faces in each video frame. A frame with larger and more faces around frame center is considered to be more attractive.
4. **Camera motion model.** Camera motion model plays a role of magnifier, which is multiplied with the sum of other visual attention models. Attention scores in zooming, dollying, panning and other camera motions are defined in an ad-hoc manner to generate an attention curve.
5. **Aural attention model.** Inspired by the observation that human are often attracted by loud or sudden sounds if they have no subjective intentions, an aural attention model is built by considering the loudness strength and variation. A sliding window is employed to compute aural

saliency along time axis. Similar to camera motion model, aural attention model also plays a role of magnifier.

6. **Speech and music attention models**. Human usually pay more attention to speech and music because they convey more semantic information than other sounds. Thus the saliency of speech/music is measured by the ratio of speech/music to other kinds of sound in a sliding window, while the sound category (i.e., speech, music, silence and others) of each small audio segment is determined by using a support vector machine. Speech and music attention curves are then generated.

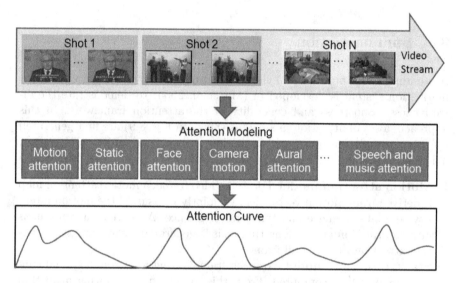

Fig. 8.9 Flowchart of attention curve generation in (Ma et al, 2005)

Given all these attention models, a unique attention curve (as shown in Fig. 8.9) is generated by combining various attention curves. The combination can be conducted either by linear or by non-linear fusion. Given the attention curve, video can be summarized either statically or dynamically. The flowchart of the summarization process is shown in Fig. 8.10. In the summarization, key-frames and video skims are extracted around the crests of the curve, which indicate the video segments probably attract viewers' attention. In order to determine the precise position of crests, a derivative curve is generated and the peaks of wave crests are those zero-crossing points from the positive to the negative on the derivative curve.

According to the attention values of key-frames, multi-scale static abstraction may be conveniently generated by ranking these attention values. For shot-based static abstraction, key-frames within a shot are used as representative frames. The maximum attention value among all key-frames in a shot is regarded as indicator of shot importance. If there is no wave crest in a shot, the middle frame is chosen, because all frames in this shot are equal important. If only one key-frame is required for each shot, the key-frame with the maximum attention value is selected. If the total number of key-frames allowed is less than shots in a video, the key-frames with lower importance values are discarded.

Moreover, dynamic video skimming generation also become much simpler with the attention curves. Given a skimming ratio, skim segments are selected around each key-frame of shot according to this ratio. In order to guarantee skimmed video sound natural and fluid, the uncompleted speech sentences should not be extracted. Although fully semantic understanding and complex heuristic rules are not required in this process, the dynamic video summarization created by user attention model is consistent with human perception. For more details, please refer to (Ma et al, 2005).

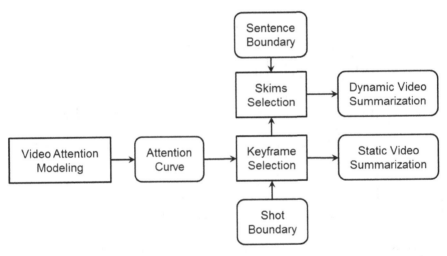

Fig. 8.10 Flowchart of the video summarization approach in (Ma et al, 2005)

8.6   Compression

Li et al (2011) proposed a saliency-based bit allocation strategy for video compression. The main idea of this approach is using the saliency map generated by the classical center-surround contrast-based saliency model to guide the bit allocation. Using a new eye-tracking-weighted evaluation metric, more than 90% of the encoded video clips with the proposed method achieve better subjective quality compared to standard encoding with comparable bit rate.

The system framework of the proposed approach is shown in Fig. 8.11. From Fig. 8.11, we can see that there are mainly two paths in the saliency-based compression. On the saliency model path (i.e., the upper half),

the current input frame is first decomposed into multi-scale analysis with
channels sensitive to low-level visual features (red-green and blue-yellow color
opponencies, temporal flicker, intensity contrast, four orientations and four
directional motion energies). The saliency map is obtained after within-
channel within-scale and cross-scale nonlinear competition. Assuming that
the top salient locations in the saliency map are likely to attract attention
and gaze of viewers, a guidance map is generated by foveating these positions.

On the compression path (i.e., the lower half), the current macro-blocks
(MBs) are predicted by previous encoded frame MBs through intra (which
means the prediction result is generated from the current frame) or inter
(which means the prediction result is generated from the previous frame)
mode. The prediction error (difference) is then passed through transform
and quantization; here the generated guidance map is used to adjust the
quantization parameters to realize the non-uniform bit allocation. Finally, an
encoded frame is complete after quantization and entropy encoding.

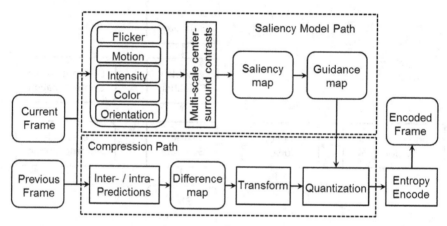

Fig. 8.11 System framework of the compression approach in (Li et al, 2011)

8.7 Notes

This Chapter introduces several prevalent saliency-based applications, while
saliency can be also used in many other scenarios. For example, Zhu et al
(2007) adopted the multi-scale structural saliency for signature detection.
López et al (2006) presented an approach to detect/track salient targets in
surveillance video. Mahadevan and Vasconcelos (2008) proposed to remove
the background from highly dynamic scenes by using a saliency model. In
(Chalmond et al, 2006), visual saliency was used to analyze huge images
obtained from remote sensing. Moreover, visual saliency can help to extract invariant features (Li and Clark, 2005), improve the performance of
scene classification (Siagian and Itti, 2007) and detect video fingerprints

(Yang et al, 2009). In these applications, visual saliency can help to allocate limited computational resource to important visual content for efficient analysis, and the analysis results can well meet human perception by focusing on the same salient subsets as human vision system does.

Generally speaking, the main objective of incorporating visual saliency is to improve user experience (e.g., enhancing the viewing experience in retargeting, reducing the intrusiveness in advertising, and summarizing attractive video contents). Note that the objective performance of these applications after considering visual saliency may only change slightly. However, the subjective evaluation may change remarkably since saliency-based analysis can better meet human perception. For instance, traditional image retrieval algorithm may return N positives in the first M results before using visual saliency. By considering visual saliency, the number of positives in the first M results may stay unchanged, but with totally different ranks. In this manner, the objective recall and precision remain unchanged, but the user experience may be better with the saliency-based re-ranking. Therefore, subjective factors should be incorporated to correctly evaluate the performance of saliency-based applications.

References

Avidan, S., Shamir, A.: Seam carving for content-aware image resizing. In: ACM SIGGRAPH. ACM, New York (2007), doi:10.1145/1275808.1276390

Chalmond, B., Francesconi, B., Herbin, S.: Using hidden scale for salient object detection. IEEE Transactions on Image Processing 15(9), 2644–2656 (2006), doi:10.1109/TIP.2006.877380

Chang, C.H., Hsieh, K.Y., Chung, M.C., Wu, J.L.: Visa: Virtual spotlighted advertising. In: Proceedings of the 16th ACM International Conference on Multimedia, MULTIMEDIA 2008, pp. 837–840. ACM, New York (2008), doi:10.1145/1459359.1459500

Elazary, L., Itti, L.: Interesting objects are visually salient. Journal of Vision 8(3), 3, 1–15 (2008), doi:10.1167/8.3.3

Fu, H., Chi, Z., Feng, D.: Attention-driven image interpretation with application to image retrieval. Pattern Recognition 39(9), 1604–1621 (2006), doi:10.1016/j.patcog.2005.12.015

Gao, W., Tian, Y., Huang, T., Yang, Q.: Vlogging: A survey of videoblogging technology on the web. ACM Computing Surveys 42(4), 15, 1–57 (2010), doi:10.1145/1749603.1749606

Guo, J., Mei, T., Liu, F., Hua, X.S.: Adon: An intelligent overlay video advertising system. In: Proceedings of the 32nd International ACM SIGIR Conference on Research and Development in Information Retrieval, SIGIR 2009, pp. 628–629. ACM, New York (2009), doi:10.1145/1571941.1572049

Lekakos, G., Papakiriakopoulos, D., Chorianopoulos, K.: An integrated approach to interactive and personalized tv advertising. In: Proceedings of Workshop on Personalization in Future TV (2001)

Li, H., Edwards, S.M., Hyun Lee, J.: Measuring the intrusiveness of advertisements: Scale development and validation. Journal of Advertising 31(2), 37–47 (2002)

Li, M., Clark, J.: Selective attention in the learning of invariant representation of objects. In: Preceedings of the IEEE Conference on Computer Vision and Pattern Recognition (CVPR) - Workshops, pp. 93–93 (2005), doi:10.1109/CVPR.2005.522

Li, S., Lee, M.C.: Efficient spatiotemporal-attention-driven shot matching. In: Proceedings of the 15th Annual ACM International Conference on Multimedia, MULTIMEDIA 2007, pp. 178–187. ACM, New York (2007), doi:10.1145/1291233.1291275

Li, Y., Wan, K.W., Yan, X., Xu, C.: Real time advertisement insertion in baseball video based on advertisement effect. In: Proceedings of the 13th Annual ACM International Conference on Multimedia, MULTIMEDIA 2005, pp. 343–346. ACM, New York (2005), doi:10.1145/1101149.1101221

Li, Y., Tian, Y., Yang, J., Duan, L.Y., Gao, W.: Video retargeting with multi-scale trajectory optimization. In: Proceedings of the International Conference on Multimedia Information Retrieval, MIR 2010, pp. 45–54. ACM, New York (2010), doi:10.1145/1743384.1743399

Li, Z., Qin, S., Itti, L.: Visual attention guided bit allocation in video compression. Image Vision Computing 29(1), 1–14 (2011), doi:10.1016/j.imavis.2010.07.001

López, M.T., Fernández-Caballero, A., Fernández, M.A., Mira, J., Delgado, A.E.: Visual surveillance by dynamic visual attention method. Pattern Recognition 39(11), 2194–2211 (2006), doi:10.1016/j.patcog.2006.04.018

Ma, Y.F., Hua, X.S., Lu, L., Zhang, H.J.: A generic framework of user attention model and its application in video summarization. IEEE Transactions on Multimedia 7(5), 907–919 (2005), doi:10.1109/TMM.2005.854410

Mahadevan, V., Vasconcelos, N.: Background subtraction in highly dynamic scenes. In: Preceedings of the IEEE Conference on Computer Vision and Pattern Recognition, CVPR, pp. 1–6 (2008), doi:10.1109/CVPR.2008.4587576

McCoy, S., Everard, A., Polak, P., Galletta, D.F.: The effects of online advertising. ACM Communication 50(3), 84–88 (2007a), doi:10.1145/1226736.1226740

McCoy, S., Everard, A., Polak, P., Galletta, D.F.: Online ad intrusiveness. In: Jacko, J.A. (ed.) HCI 2007. LNCS, vol. 4553, pp. 86–89. Springer, Heidelberg (2007)

Mei, T., Hua, X.S., Yang, L., Li, S.: Videosense: Towards effective online video advertising. In: Proceedings of the 15th Annual ACM International Conference on Multimedia, MULTIMEDIA 2007, pp. 1075–1084. ACM, New York (2007), doi:10.1145/1291233.1291467

Siagian, C., Itti, L.: Rapid biologically-inspired scene classification using features shared with visual attention. IEEE Transactions on Pattern Analysis and Machine Intelligence 29(2), 300–312 (2007), doi:10.1109/TPAMI.2007.40

Thawani, A., Gopalan, S., Sridhar, V.: Context aware personalized ad insertion in an interactive tv environment. In: Proceedings of Workshop on Personalization in Future TV (2004)

Wang, J., Fang, Y., Lu, H.: Online video advertising based on user's attention relavancy computing. In: Preceedings of the IEEE International Conference on Multimedia and Expo, ICME, pp. 1161–1164 (2008), doi:10.1109/ICME.2008.4607646

Yang, R., Tian, Y., Huang, T.: Dct-based videoprinting on saliency-consistent regions for detecting video copies with text insertion. In: Muneesawang, P., Wu, F., Kumazawa, I., Roeksabutr, A., Liao, M., Tang, X. (eds.) PCM 2009. LNCS, vol. 5879, pp. 797–806. Springer, Heidelberg (2009)

Zhu, G., Zheng, Y., Doermann, D., Jaeger, S.: Multi-scale structural saliency for signature detection. In: Preceedings of the IEEE Conference on Computer Vision and Pattern Recognition, CVPR, pp. 1–8 (2007), doi:10.1109/CVPR.2007.383255

Chapter 9
Conclusions and Future Work

In this book, we have introduced the efforts in visual saliency computation from multiple aspects, including the key findings from neurobiology and psychology, location-based and object-based saliency, bottom-up and top-down models, evaluation methodologies and applications. Note that we ignore many tedious technical details and try to make the content easy to understand. We believe this book can serve as a good start for those who are interested in visual saliency computation.

In this chapter, we first summarize our contributions in Sect. 9.1. After that, we discuss future works in visual saliency computation in Sect. 9.2.

9.1 Contribution Summarization

The main contributions of the book are summarized as follows:

1. We introduce the key neurobiological and psychological findings that are believed to be related with visual saliency. In particular, we omit the technical details and only present the main findings from neurobiology and psychology to support the computational modeling of visual saliency. Thus this survey can serve as a good reference for non-professionals.
2. A systematic survey of bottom-up saliency models is presented to show the main branches in computing location-based saliency. Moreover, a complete overview of benchmarks and evaluation metrics is also presented to quantize the performance of saliency models. With these materials, readers can easily understand how to build and evaluate visual saliency models.
3. We also introduce several representative models for object-based saliency computation. In particular, we have proposed two novel approaches to detect salient objects. The first approach utilizes the complementary location-based saliency maps to segment salient objects, while the second one

directly computes object-based saliency through single image optimization. Comparisons with existing approaches show that these two approaches outperform state-of-the-art approaches.

4. To integrate bottom-up and top-down factors, we present a probabilistic multi-task learning approach and a statistical learning approach for visual saliency computation. In these two approaches, the bottom-up and top-down factors are considered simultaneously in a probabilistic/Bayesian framework. In this framework, a bottom-up component simulates the low-level processes in human vision system; while a top-down component simulates the high-level processes to bias the competition of the input visual stimuli. Extensive experiments show that these approaches demonstrate high robustness and effectiveness in computing visual saliency.

5. For the problem of mining cluster-specific top-down priors, this book presents a multi-task rank learning approach for visual saliency computation. In this approach, the problem of visual saliency computation is formulated in a learning to rank framework to infer multiple saliency models that apply to different scene clusters. In the training process, this approach can infer multiple visual saliency models simultaneously. With an appropriate sharing of information across models, the generalization ability of each model can be greatly improved. Extensive experiments on the eye-fixation dataset show that this approach is highly effective in computing visual saliency in various categories of video scenes.

6. Two novel approaches are proposed to remove the label ambiguities in user data when training saliency models, including the cost-sensitive learning to rank approach and the multi-instance learning to rank approach. These two approaches focus on removing the label ambiguities by using the pairwise correlations between training instances. Experimental results prove that visual saliency models can achieve impressive improvements by focusing on the correlations between real targets and real distractors.

7. This book also investigates the usage of visual saliency models in computer vision and multimedia applications. Objective and subjective evaluation shows that visual saliency model can help to increase the efficiency and obtain impressive results to meet human perception by focusing on the same salient visual subsets as human vision system does.

From these contributions, we can see the main highlights of this book are from three aspects: 1) systematic studies on the neurobiological/psychological findings, saliency benchmarks, models and applications; 2) learning-based approaches to handle the top-down factors in visual saliency computation; and 3) innovative approaches for detecting salient objects. Among these highlights, the following approaches are first proposed and published by authors of the book:

1. **Benchmark**. In Sect. 2.2.2, the video benchmark with all salient objects in the key frames labeled by multiple subjects with rectangles was proposed in (Li et al, 2009).

2. **Object-based saliency**. In Sect. 4.2.2, the approach to detect salient objects from complementary location-based saliency maps was proposed in (Yu et al, 2010). The approach in Sect. 4.3.4 which aimed to optimize object-based saliency through single image optimization was proposed in (Li et al, 2013a).

3. **The probabilistic multi-task learning approach**. The probabilistic multi-task learning approach in Chap. 5 was proposed in (Li et al, 2010b), which focused on the integration of bottom-up and top-down factors.

4. **The statistical learning approach**. The saliency model that utilized the statistical priors learned from millions of images in Chap. 5 was proposed in (Li et al, 2013b), which focused on modulating the bottom-up saliency with the learned statistical priors.

5. **Cluster-specific model**. In Chap. 6, the multi-task learning to rank approach was proposed in (Li et al, 2011) to mine the cluster-specific knowledge for top-down modulation.

6. **Label ambiguity removal**. In Chap. 7, we present two approaches to remove the label ambiguities in user data when training saliency models. These two approaches were first proposed in (Li et al, 2010a) and (Li et al, 2012), respectively.

7. **Video Retargeting**. In Sect. 8.2.2, the optimization-based video retargeting approach was first proposed in (Li et al, 2010c).

In conclusion, this book summarizes various aspects of visual saliency computation, from the theoretical foundations to real-world applications. In particular, this book proposes that machine learning algorithms can be extremely helpful in solving several key problems in computing visual saliency, while these problems cannot be easily addressed by directly simulating the bottom-up saliency mechanisms in human vision system. To prove this, this book presents several innovative studies on how to apply machine learning algorithms into visual saliency computation. In these studies, the feasibility and effectiveness of learning-based visual saliency models are demonstrated in extensive experiments. We believe that such results will spark a great interest of research in the related communities in years to come.

9.2   Future Work

From the discussions above, we find that visual saliency computation remains an open problem and some challenges still need to be addressed. For these challenges, we propose three possible research directions:

1. **Benchmarking**. Existing benchmarks on visual saliency only contain limited scenes, especially for the image benchmarks. When using these limited scenes in the evaluation, the performance of the testing model can be revealed to some extent. However, the small benchmark lacks the ability

to explore the fundamental mechanism of visual saliency computation in general cases. For example, a benchmark with outdoor images may reveal how subjects react when encountering natural scenes. However, models fine-tuned on this benchmark may fail to process indoor scenes and artificial images (e.g., cartoon and news). Therefore, it is necessary to build a large benchmark that covers massive possible scenes to explore the most effective mechanisms in visual saliency computation.

2. **Learning objective.** Modeling the top-down factors in human vision system requires to learn the probable modulation mechanisms hidden in user data. Actually, the learning objective can be described as various kinds of subjective bias people may apply in different scenarios. For example, the feature bias can be modeled by finding the optimal strategies to combine visual features, while the location bias is usually modeled by the 2D Gaussian re-weighting around image center. However, there still exist many other kinds of bias that should be learned. For example, various parts of a complex object (e.g., a car) may be inherently correlated and should be processed as a whole in visual saliency computation even if they have distinct visual appearances. Since it is difficult to directly infer such latent correlations from the input scene, it is necessary to explore how to learn them from training data.

3. **Learning algorithms.** From the learning algorithms used in this book, we can see that there are at least three supervised ways to learning top-down factors, including classification, regression and ranking. Classification aims to directly identify whether a subset is salient or not, while regression aims to optimize "stimulus-saliency" mapping functions to obtain the saliency value. In experiments, we find that learning to rank algorithms, which only focus whether a subset is more salient than the other one, have achieved the best performance. Actually, the ranking formulation is compatible with mechanisms in human vision system which turn the parallel information into serial processing according to their priority. Therefore, the ranking framework could be a good candidate for designing learning algorithms. Meanwhile, the unsupervised learning algorithms should also be also considered since there are much more unlabeled data. For example, such unsupervised learning algorithms can be used to mine the latent signal correlations from millions of unlabeled images. Therefore, it is necessary to combine the unsupervised learning methodology and supervised ranking framework to better model the top-down factors.

Beyond these three directions, there are some other feasible solutions to improve the visual saliency models, e.g., incorporating high-level features such as face and object category, mining the non-linear fusion strategies between bottom-up and top-down factors, and exploring compact signal representations. We believe these attempts on large-scale benchmarks will progressively improve the performance of computer vision system, which, in the future, may help us to further explore the unknown mechanisms in human vision system.

References

Li, J., Tian, Y., Huang, T., Gao, W.: A dataset and evaluation methodology for visual saliency in video. In: Preceedings of the IEEE International Conference on Multimedia and Expo (ICME), pp. 442–445 (2009), doi:10.1109/ICME.2009.5202529

Li, J., Tian, Y., Huang, T., Gao, W.: Cost-sensitive rank learning from positive and unlabeled data for visual saliency estimation. IEEE Signal Processing Letters 17(6), 591–594 (2010a), doi:10.1109/LSP.2010.2048049

Li, J., Tian, Y., Huang, T., Gao, W.: Probabilistic multi-task learning for visual saliency estimation in video. International Journal of Computer Vision 90(2), 150–165 (2010b), doi:10.1007/s11263-010-0354-6

Li, J., Tian, Y., Huang, T., Gao, W.: Multi-task rank learning for visual saliency estimation. IEEE Transactions on Circuits and Systems for Video Technology 21(5), 623–636 (2011), doi:10.1109/TCSVT.2011.2129430

Li, J., Xu, D., Gao, W.: Removing label ambiguity in learning-based visual saliency estimation. IEEE Transactions on Image Processing 21(4), 1513–1525 (2012), doi:10.1109/TIP.2011.2179665

Li, J., Tian, Y., Duan, L., Huang, T.: Estimating visual saliency through single image optimization. IEEE Signal Processing Letters 20(9), 845–848 (2013a), doi:10.1109/LSP.2013.2268868

Li, J., Tian, Y., Huang, T.: Visual saliency with statistical priors. International Journal of Computer Vision, 1–15 (2013b), doi:10.1007/s11263-013-0678-0

Li, Y., Tian, Y., Yang, J., Duan, L.Y., Gao, W.: Video retargeting with multi-scale trajectory optimization. In: Proceedings of the International Conference on Multimedia Information Retrieval (MIR 2010), pp. 45–54. ACM, New York (2010c), doi:10.1145/1743384.1743399

Yu, H., Li, J., Tian, Y., Huang, T.: Automatic interesting object extraction from images using complementary saliency maps. In: Proceedings of the International Conference on Multimedia (MULTIMEDIA 2010), pp. 891–894. ACM, New York (2010), doi:10.1145/1873951.1874105

Index